SAFETY AND RISK IN SOCIETY

CHEMICAL FACILITY SECURITY: CONSIDERATIONS FOR CONGRESS

SAFETY AND RISK IN SOCIETY

Additional books in this series can be found on Nova's website under the Series tab.

Additional E-books in this series can be found on Nova's website under the E-book tab.

DEFENSE, SECURITY AND STRATEGIES

Additional books in this series can be found on Nova's website under the Series tab.

Additional E-books in this series can be found on Nova's website under the E-book tab.

SAFETY AND RISK IN SOCIETY

CHEMICAL FACILITY SECURITY: CONSIDERATIONS FOR CONGRESS

MARLIN J. FLORES
AND
NICHOLAS J. WASHINGTON
EDITORS

Nova Science Publishers, Inc.
New York

For permission to use material from this book please contact us:
Telephone 631-231-7269; Fax 631-231-8175
Web Site: http://www.novapublishers.com

Library of Congress Cataloging-in-Publication Data

ISBN 978-1-61942-925-3

Published by Nova Science Publishers, Inc. †New York

CONTENTS

PREFACE

Even before September 11, 2001, congressional policymakers have expressed concern about the safety and security of facilities possessing certain amounts of hazardous chemicals. The sudden release of hazardous chemicals from facilities storing large quantities might potentially harm many people living or working near the facility. Chemical facilities engaged in security activities on a voluntary basis. Following September 11, 2001 some states enacted laws requiring additional consideration of security at chemical facilities. Congress debated whether the federal government should regulate such facilities for security purposes to reduce the risk they pose. This book provides a brief overview of the existing statutory authority to regulate chemical facilities with a focus on policy issues and options for congressional consideration.

Chapter 1 - The Department of Homeland Security (DHS) has statutory authority to regulate chemical facilities for security purposes. The 112th Congress extended this authority through November 18, 2011. The 112th Congress debated the scope and details of reauthorization and continues to consider legislation establishing an authority with longer duration. Some Members of Congress support an extension, either short- or long-term, of the existing authority. Other Members call for revision and more extensive codification of chemical facility security regulatory provisions. Questions regarding the current law's effectiveness in reducing chemical facility risk and the sufficiency of federal funding for chemical facility security exacerbate the tension between continuing current policies and changing the statutory authority.

The DHS is in the process of implementing the authorized regulations, called chemical facility anti-terrorism standards (CFATS). The DHS finalized

CFATS regulations in 2007. No chemical facilities have completed the CFATS process, which starts with information submission by chemical facilities and finishes with inspection and approval of facility security measures by DHS. Several factors, including the level of detail provided to DHS and the availability of inspectors, likely complicate the inspection process and lead to delays in inspection. Policymakers have questioned whether the compliance rate with the CFATS is sufficient to address this homeland security issue.

Key policy issues debated in previous Congresses contribute to the reauthorization debate. These issues include the adequacy of DHS resources and effort; the appropriateness and scope of federal preemption of state chemical facility security activities; the availability of information for public comment, potential litigation, and congressional oversight; the universe of facilities that are considered as chemical facilities; and the role of inherently safer technologies in achieving security goals.

The 112th Congress might take various approaches to this issue. Congress might allow the statutory authority to expire but continue providing appropriations to administer the regulations. Congress might permanently or temporarily extend the expiring statutory authority in order to observe the impact of the current regulations and, if necessary, address any perceived weaknesses at a later date. Congress might codify the existing regulation in statute and reduce the discretion available to the Secretary of Homeland Security to change the current regulatory framework. Alternatively, Congress might substantively change the current regulation's implementation, scope, or impact by amending the existing statute or creating a new one. Finally, Congress might choose to terminate the program by allowing its authority to lapse and removing funding for the program. This last approach would leave chemical facility security regulation to the discretion of state and local governments.

Both authorization and appropriation legislation in the 112th Congress addresses chemical facility security. The Senate- and House-passed continuing resolution, H.R. 2608, would extends the existing authority until November 18, 2011. Authorizing legislation in the House includes H.R. 225; H.R. 901, ordered reported as amended by the House Committee on Homeland Security and referred to the House Committee on Energy and Commerce; H.R. 908, ordered reported as amended by the House Committee on Energy and Commerce; H.R. 916; H.R. 2890; S. 473, ordered reported as amended by the Senate Committee on Homeland Security and Governmental Affairs, and S. 709.

Chapter 2 - Dan Lungren welcomes witnesses to the first Subcommittee hearing under the new Republican majority. The subcommittee will be examining many critical issues in this Congress from physical and cyber threats to our critical infrastructure, to radinuc/bio threats to our cities. They will be aggressive in their oversight of the department and its many security programs, especially those that provide "no substantial security improvement" for the amount of taxpayer dollars spent.

Chapter 3 - Rand Beers Thanks Chairman Lungren, Ranking Member Clarke, and distinguished members of the subcommittee. The department's current authority under Section 550 of the Fiscal Year 2007 Department of Homeland Security Appropriations Act, as amended, was set to expire in October 2010, but has been temporarily extended under the current Continuing Resolution. DHS is eager to work with this committee, Congress, and all levels of government and the private sector to achieve passage of legislation that permanently authorizes and appropriately matures the Chemical Facility Anti- Terrorism Standards (CFATS) program. In the interest of facilitating that collaboration, Beers's testimony focuses on the current program and the key principles that DHS would like to see guide the program's maturation.

Chapter 4 - Timothy Scott, Chief Security Officer for the Dow Chemical Company (Dow is a member of the American Chemistry Council) speaks on behalf of Dow and Dow's industry association. He makes three key points in his statement that demonstrate that Dow has a strong foundation in place to build from when it comes to securing the nation's chemical facilities:

First – The chemical industry is critical to our national economy as well as the quality of life of our people. We recognize the historical and current risks to our industry and have taken aggressive action to improve our security posture – voluntarily well before the creation of DHS and the Chemical Facility Anti-Terrorism Standards (CFATS) and continuing now with the ongoing implementation of CFATS.

Second - The implementation of CFATS to date is demonstrably achieving its objectives to reduce the number of high risk sites and lower the risk profile of remaining high risk sites.

And third – and most important – we need to continue on this successful path, maintain the momentum we've achieved and complete the task at hand without further delays or calls for change that would result in slowing the progress being made on chemical facility security by both DHS and the chemical industry..

Chapter 5 - M. Sam Mannan holds a BS, MS, and PhD in chemical engineering. He is a registered professional engineer in the states of Louisiana and Texas and a certified safety professional. Dr. Mannan is a Fellow of the American Institute of Chemical Engineers and a member of the American Society of Safety Engineers, the International Institute of Ammonia Refrigeration, and the National Fire Protection Association. He is Director of the Mary Kay O'Connor Process Safety Center, holder of the T. Michael O'Connor Chair I in Chemical Engineering, and Regents Professor of Chemical Engineering at Texas A&M University. The center seeks to develop safer processes, equipment, procedures, and management strategies that will minimize losses in the process industry. Mannan's area of expertise within the chemical engineering discipline is process safety. He teaches process safety engineering both at the undergraduate and graduate level. He also teaches continuing education courses on process safety and other specialty process safety courses in the United States and overseas. His research and practice is primarily in the area of process safety and related subjects. The opinions presented in this document represent my personal position on these issues. These opinions are based on my education, experience, research and training.

Chapter 6 - George Hawkins is the General Manager of the District of Columbia Water and Sewer Authority – also known simply as DC Water. He testifies on the circumstances surrounding the decision to have the Blue Plains Wastewater Treatment Facility voluntarily switch from using a chlorine in the treatment of wastewater to a potentially safer alternative.

Chapter 7 - Peter Silva is Assistant Administrator for Water at the United States Environmental Protection Agency. He welcomes discusses the EPA's efforts to promote security and resiliency in the Water Sector with an emphasis on their role in addressing chemical security at water facilities.

In: Chemical Facility Security
Editors: M. J. Flores et al.
ISBN 978-1-61942-925-3

Chapter 1

CHEMICAL FACILITY SECURITY: ISSUES AND OPTIONS FOR THE 112TH CONGRESS*

Dana A. Shea

SUMMARY

The Department of Homeland Security (DHS) has statutory authority to regulate chemical facilities for security purposes. The 112th Congress extended this authority through November 18, 2011. The 112th Congress debated the scope and details of reauthorization and continues to consider legislation establishing an authority with longer duration. Some Members of Congress support an extension, either short- or long-term, of the existing authority. Other Members call for revision and more extensive codification of chemical facility security regulatory provisions. Questions regarding the current law's effectiveness in reducing chemical facility risk and the sufficiency of federal funding for chemical facility security exacerbate the tension between continuing current policies and changing the statutory authority.

The DHS is in the process of implementing the authorized regulations, called chemical facility anti-terrorism standards (CFATS). The DHS finalized CFATS regulations in 2007. No chemical facilities have completed the

* This is an edited, reformatted and augmented version of the Congressional Research Service Publication, R41642, dated October 7, 2011.

CFATS process, which starts with information submission by chemical facilities and finishes with inspection and approval of facility security measures by DHS. Several factors, including the level of detail provided to DHS and the availability of inspectors, likely complicate the inspection process and lead to delays in inspection. Policymakers have questioned whether the compliance rate with the CFATS is sufficient to address this homeland security issue.

Key policy issues debated in previous Congresses contribute to the reauthorization debate. These issues include the adequacy of DHS resources and effort; the appropriateness and scope of federal preemption of state chemical facility security activities; the availability of information for public comment, potential litigation, and congressional oversight; the universe of facilities that are considered as chemical facilities; and the role of inherently safer technologies in achieving security goals.

The 112th Congress might take various approaches to this issue. Congress might allow the statutory authority to expire but continue providing appropriations to administer the regulations. Congress might permanently or temporarily extend the expiring statutory authority in order to observe the impact of the current regulations and, if necessary, address any perceived weaknesses at a later date. Congress might codify the existing regulation in statute and reduce the discretion available to the Secretary of Homeland Security to change the current regulatory framework. Alternatively, Congress might substantively change the current regulation's implementation, scope, or impact by amending the existing statute or creating a new one. Finally, Congress might choose to terminate the program by allowing its authority to lapse and removing funding for the program. This last approach would leave chemical facility security regulation to the discretion of state and local governments.

Both authorization and appropriation legislation in the 112th Congress addresses chemical facility security. The Senate- and House-passed continuing resolution, H.R. 2608, would extends the existing authority until November 18, 2011. Authorizing legislation in the House includes H.R. 225; H.R. 901, ordered reported as amended by the House Committee on Homeland Security and referred to the House Committee on Energy and Commerce; H.R. 908, ordered reported as amended by the House Committee on Energy and Commerce; H.R. 916; H.R. 2890; S. 473, ordered reported as amended by the Senate Committee on Homeland Security and Governmental Affairs, and S. 709.

INTRODUCTION

Even before September 11, 2001, congressional policymakers have expressed concern about the safety and security of facilities possessing certain amounts of hazardous chemicals. The sudden release of hazardous chemicals from facilities storing large quantities might potentially harm many people living or working near the facility. Chemical facilities engaged in security activities on a voluntary basis. Following September 11, 2001, some states enacted laws requiring additional consideration of security at chemical facilities.[1] Congress debated whether the federal government should regulate such facilities for security purposes to reduce the risk they pose. The 109th Congress passed legislation in 2006 providing the Department of Homeland Security (DHS) statutory authority to regulate chemical facilities for security purposes, and subsequent Congresses have extended this authority. The statutory authority expires November 18, 2011. Advocacy groups, stakeholders, and policymakers have called for congressional reauthorization of this authority, though they disagree about the preferred approach. Congress may extend the existing authority, revise the existing authority to resolve potentially contentious issues, or allow this authority to lapse.

This report provides a brief overview of the existing statutory authority and the regulation implementing this authority. It describes several policy issues raised in previous debates regarding chemical facility security and identifies policy options for congressional consideration. Finally, legislation in the 112th Congress is discussed.

OVERVIEW OF STATUTE AND REGULATION

The 109th Congress provided DHS with statutory authority to regulate chemical facilities for security purposes.[2] The statute explicitly identified some DHS authorities and left other aspects to the discretion of the Secretary of Homeland Security. The statute contains a "sunset provision" and expires on November 18, 2011.[3]

On April 9, 2007, the Department of Homeland Security issued an interim final rule regarding the chemical facility anti-terrorism standards (CFATS).[4] This interim final rule entered into force on June 8, 2007. The interim final rule implements both statutory authority explicit in P.L. 109-295, Section 550, and authorities DHS found Congress implicitly granted. In promulgating the interim final rule, DHS interpreted the language of the statute to determine what it asserts was the intent of Congress when crafting the statutory authority. Consequently, much of the rule arises from the Secretary's discretion and interpretation of legislative intent rather than explicit statutory language.

Under the interim final rule, the Secretary of Homeland Security determines which chemical facilities must meet regulatory security requirements, based on the degree of risk posed by each facility. The DHS lists 322 chemicals as "chemicals of interest" for the purposes of compliance with CFATS.[5] The DHS considers each chemical in the context of three threats: release; theft or diversion; and sabotage and contamination. Chemical facilities with greater than specified quantities of potentially dangerous chemicals must submit information to DHS, so that DHS can determine the facility's risk status. The statute exempts several types of facilities from this requirement: facilities defined as a water system or wastewater treatment works; facilities owned or operated by the Department of Defense or Department of Energy; facilities regulated by the Nuclear Regulatory Commission; and those facilities regulated under the Maritime Transportation Security Act of 2002 (P.L. 107-295).

Based on the submitted information, DHS determines the risk associated with each facility. Facilities DHS deems high risk must meet CFATS requirements. The DHS assigns high-risk facilities into one of four risk-based tiers. The statute mandated that DHS establish performancebased security requirements.[6] The DHS created graduated performance-based requirements for facilities assigned to each risk-based tier. Facilities in higher risk tiers must meet more stringent performance-based requirements.

All high-risk facilities must assess their vulnerabilities, develop an effective security plan, submit these documents to DHS, and implement their security plan.[7] The vulnerability assessment serves two purposes under the interim final rule. One is to determine or confirm the placement of the facility in a risk-based tier. The other is to provide a baseline against which to evaluate the site security plan activities.

The site security plans must address the vulnerability assessment by describing how activities in the plan correspond to securing facility vulnerabilities. Additionally, the site security plan must address preparations for and deterrents against specific modes of potential terrorist attack, as applicable and identified by DHS. The site security plans must also describe how the activities taken by the facility meet the risk-based performance standards provided by DHS.

The DHS must review and approve the submitted documents, audit and inspect chemical facilities, and determine regulatory compliance. The DHS may disapprove submitted vulnerability assessments or site security plans that fail to meet DHS standards. The DHS may not disapprove an assessment or plan because of the presence or absence of a specific security measure. In the case of disapproval, DHS must identify in writing those areas of the assessment and/or plan that need improvement. Owners or operators of chemical facilities may appeal such decisions to DHS.

Similarly, if, after inspecting a chemical facility, DHS finds a facility not in compliance, the Secretary must write to the facility explaining the deficiencies found, provide an opportunity for the facility to consult with DHS, and issue an order to the facility to comply by a specified date. If the facility continues to be out of compliance, DHS may fine and, eventually, order the facility to cease operation. The interim final rule establishes the process by which chemical facilities can appeal DHS decisions and rulings, but the statute prohibits third-party suits for enforcement purposes.

The interim final rule creates a category of information exempted from disclosure under the Freedom of Information Act (FOIA) and comparable state and local laws. The DHS named this category of information "Chemical-terrorism Vulnerability Information" (CVI). This category exempts information generated under the interim final rule, as well as any information developed for chemical facility security purposes identified by the Secretary. In accordance with statute, judicial and administrative proceedings shall treat CVI as classified information. The DHS asserts sole discretion regarding who will be eligible to receive CVI. Disclosure of CVI may be punishable by fine.

The interim final rule states it preempts state and local regulation that "conflicts with, hinders, poses an obstacle to or frustrates the purposes of" the federal regulation. States, localities, or affected companies may request a decision from DHS regarding potential conflict between the regulations. Since DHS promulgated the interim final rule, Congress amended P.L. 109-295, Section 550, to state that such preemption will occur only in the case of an "actual conflict."[8] The DHS has not issued revised regulations addressing this change in statute.

IMPLEMENTATION

Within DHS, the National Protection and Programs Directorate (NPPD) is responsible for chemical facility security regulations. The NPPD generally attempts to reduce the risks to the homeland and has various offices addressing both physical and virtual threats. Within NPPD, the Office of Infrastructure Protection, through its Infrastructure Security Compliance Division, oversees the CFATS program.[9] As seen in **Table 1**, requested and appropriated funding for this program has generally increased since its creation. Additionally, full-time equivalent staffing for this program has also generally increased. This increase in staffing reflects, in part, the development of a cadre of CFATS inspectors, based in regional offices.

Table 1. DHS Funding for Chemical Facility Security Regulation by Fiscal Year

Fiscal Year	Request ($ in millions)	Appropriation ($ in millions)	Full-time Equivalents
FY2007	10	22a	0
FY2008	25	50	21
FY2009	63	78b	78
FY2010	103c	103d	246
FY2011	105e	96f	257
FY2012	99g		242

Source: Department of Homeland Security, Preparedness Directorate, Infrastructure Protection and Information Security, *FY2007 Congressional Justification*; Department of Homeland Security, National Protection and Programs Directorate, Infrastructure Protection and Information Security, *Fiscal Year 2008 Congressional Justification*; Department of Homeland Security, National Protection and Programs Directorate, Infrastructure Protection and Information Security, *Fiscal Year 2009 Congressional Justification*; Department of Homeland Security, National Protection and Programs Directorate, Infrastructure Protection and Information Security, *Fiscal Year 2010 Congressional Justification*; Department of Homeland Security, National Protection and Programs Directorate, Infrastructure Protection and Information Security, *Fiscal Year 2011 Overview Congressional Justification*; Department of Homeland Security, National Protection and Programs Directorate, Infrastructure Protection and Information Security, *Fiscal Year 2012 Congressional Justification*; H.Rept. 109-699; P.L. 110-28; the explanatory statement for P.L. 110-161 *at Congressional Record*, December 17, 2007, p. H16092; the explanatory statement for P.L. 110-329 *at Congressional Record*, September 24, 2008, pp. H9806-H9807; H.Rept. 111-298; P.L. 111-242, as amended; and P.L. 112-10.

Notes: Funding levels rounded to nearest million. A full-time equivalent equals one staff person working a fulltime work schedule for one year.

a. Including funds provided in supplemental appropriations (P.L. 110-28).

b. Of this amount appropriated for the Infrastructure Security Compliance Project, $5 million were designated for activities related to the development of ammonium nitrate regulations.
c. Of this amount requested for the Infrastructure Security Compliance Project, $14 million were designated for activities related to the development of ammonium nitrate regulations.
d. Of this amount appropriated for the Infrastructure Security Compliance Project, $14 million were designated for activities related to the development of ammonium nitrate regulations.
e. The DHS planned to use an unspecified amount of the requested funds to regulate ammonium nitrate sale and transfer.
f. P.L. 112-10 provided $838,763,112 in FY2011 appropriations for the National Protection and Programs Directorate Infrastructure Protection and Information Security appropriations account. It does not specify a level for chemical facility security regulation. The DHS determined the distribution of funding for programs within the account. The DHS uses an unspecified amount of these funds to regulate ammonium nitrate sale and transfer.
g. The DHS plans to use an unspecified amount of the requested funds to regulate ammonium nitrate sale and transfer.

The DHS received statutory authority to regulate chemical facilities in 2006. It did not possess a chemical facility security office or inspector cadre at that time. The DHS requested additional positions to create an inspector cadre and is still in the process of hiring. As of February 2011, DHS had hired 109 inspectors.[10]

Chemical inspectors must be able to assess the security measures at a chemical facility using the performance-based criteria developed by DHS. Performance-based security measures are likely more difficult than prescriptive measures for chemical inspectors to assess and thus require greater training and experience in the inspector cadre. In order to overcome this challenge, DHS has established a Basic Inspector School training program for its inspector cadre. Such training, while likely improving the quality of inspection, also introduces additional time between the hiring of new inspectors and their deployment in the field.

As of September 2011, more than 38,000 chemical facilities had registered with DHS and completed the Top-Screen process, the process by which DHS determines whether a facility is high-risk based on a facility's initial submission of information. Of these facilities, DHS considered more than 7,000 as preliminarily high-risk and required to submit a site vulnerability assessment.[11] From the submitted site vulnerability assessments, DHS identified and placed 4,569 facilities into preliminary or final risk tiers.

Table 2 identifies by risk tier the universe of high-risk facilities, with Tier 1 those of highest risk.

Table 2. Facilities Regulated by DHS under CFATS

Risk Tier	Facilities with Regulated Final Tier Decision	Facilities Awaiting Final Tier Decision	Total Facilities
1	99	3	102
2	502	37	539
3	1,155	135	1290
4	2,195	443	2,638
Total	**3,951**	**618**	**4,569**

Source: Personal communication with Department of Homeland Security, September 15, 2011. See also, AcuTech Consulting Group, *A Survey of CFATS Progress in Securing the Chemical Sector*, September 6, 2011.

Notes: DHS has preliminarily assigned some facilities to a risk tier. Final assignment to a risk tier occurs after final review of submitted vulnerability assessments. In June 2011, DHS reassigned approximately 500 facilities from their risk tier to a lower risk tier.

In June 2011, DHS identified an anomaly in one of the risk-assessment tools used by DHS to determine a facility's risk tier.[12] Subsequent review of this risk-assessment tool resulted in DHS reassigning approximately 500 facilities from their risk tier to a lower risk tier.[13] The DHS lowered the number of facilities allocated to the highest-risk tier from 211 to 102, a greater than 50% reduction.[14] In some cases, DHS determined that regulated facilities no longer qualified as a high-risk facility and thus did not receive a risk tier.

The total number of chemical facilities assigned a risk tier by DHS has declined since the CFATS program began. Several factors may have contributed to this decline, including erroneous filing by regulated entities, process changes on the part of regulated entities, and business operations and decisions. The DHS has also engaged in targeted outreach activities to identify those facilities that fall under the regulation but have not yet complied by filing required information.

The DHS planned to begin inspections of Tier 1 facilities as quickly as 14 months after issuance of regulations.[15] A series of factors has delayed inspections, including the release of additional regulatory information in the form of an appendix and the need to build an inspector cadre, to establish a regional infrastructure, and to perform pre-authorization inspections at facilities. DHS officials provided a series of timeframes for beginning inspections.[16] The DHS began inspections of Tier 1 facilities in February 2010.[17] Although the DHS testified that they planned to inspect all Tier 1 facilities by the end of calendar year 2010,[18] DHS had only performed nine authorization inspections as of September 2011.[19] The DHS states that it expects to inspect all Tier 1 facilities by the end of calendar year 2011. The DHS has not approved the implementation of any site security plan.[20]

The DHS identified an additional factor in the delay of the inspection schedule: the necessary iteration between DHS and the regulated entity regarding its site security plan.[21] The DHS has issued 66 administrative orders to compel facilities to complete their site security plans.[22] In addition, DHS established a pre-authorization inspection process to gain additional information from facilities in order to fully assess the submitted site security plan. Once DHS completes a pre-authorization inspection at a facility, the facility may amend its site security plan to reflect the results of the pre-authorization inspection. The DHS has performed 180 pre-authorization inspections to date.[23]

POLICY ISSUES

Previous congressional discussion on chemical facility security raised several contentious policy issues.[24] Some issues, such as whether DHS has sufficient funds to adequately oversee chemical facility security; whether federal chemical facility security regulations should preempt state regulations; and how much chemical security information individuals may share outside of the facility and the federal government, will exist even if Congress extends the existing statutory authority. Other issues, such as what facilities DHS should regulate as a chemical facility and whether DHS should require chemical facilities to adopt or consider adopting inherently safer technologies, may be more likely addressed if Congress chooses to revise or expand existing authority.

Adequacy of Funds

The regulation establishes an oversight structure that relies on DHS personnel inspecting chemical facilities and ascertaining whether regulated entities have implemented their approved site security plans. Although the use of performance-based measures, where chemical facilities have flexibility in how to achieve the required security performance, may reduce some demands on the regulated entities, it may also require greater training and judgment on the part of DHS inspectors. Inspecting the regulated facilities likely will be costly. Congressional oversight has raised the question of whether DHS has requested and received appropriated funds sufficient to hire and retain the staff necessary to perform the required compliance inspections.[25]

The DHS may face challenges when creating the necessary infrastructure to perform nationwide inspections. As stated by DHS when describing its efforts to hire, train, and deploy an inspector cadre and support staff:

> Infrastructure Security Inspectors, located in up to 10 primary field offices across the Nation, will inspect and ensure regulatory compliance at facilities covered by the CFATS regulation, including site security plan approval and maintaining respective inspection and audit schedule. Creating a fully functional cadre will require not just recruiting and training staff, but also procurement of communications and [information technology] equipment (laptops, blackberries, etc.) to facilitate work efforts while conducting inspections and traveling, but also the acquisition of office space and equipment, government vehicles, support staff, safety equipment and clothing, and support for frequent travel.[26]

The degree to which funds meet agency needs likely depends on factors external and internal to DHS. External factors include the number of regulated facilities and the sufficiency of security plan implementation. Internal factors include the ratio between headquarters staff and field inspectors; the risk tiers of the regulated facilities; and the timetable for implementation of inspections. Once the DHS determines the number of regulated facilities and their associated timetables, DHS may be able to more comprehensively determine its resource needs.[27] Now that DHS has begun implementation of these requirements, it may be able to provide further estimates of both funding and staff requirements.

Rate of Inspection

As of February 2011, no chemical facilities have completed the CFATS process, which starts with information submission by chemical facilities and finishes with inspection and approval of security measures by DHS.[28] Some policymakers have expressed surprise at the pace of inspection and questioned whether DHS should continue at the current pace or accelerate the compliance process.[29] Several factors likely complicate the inspection process and lead to delays in inspection. A primary factor appears to be that the information facilities submit in site security plans does not provide what DHS views as necessary detail to evaluate compliance.[30] Rather than reject these site security plans, DHS has implemented an additional inspection function, a preauthorization inspection, to allow DHS to gather the necessary information from regulated facilities.

While pre-authorization inspections may lead to higher quality site security plan submissions, they appear to be a significant drain on DHS resources. The DHS cites it has performed 180 preauthorization inspections but only nine authorization inspections.[31] In principle, such preauthorization inspections may lower the future authorization inspection burden, as CFATS inspectors will be familiar with security measures at the chemical facility. Such familiarity may hasten the actual authorization inspection.

The DHS has also suggested that pre-authorization inspections are most necessary at higher risk tier facilities, due to the complexity of the facility, the potential presence of multiple chemicals of interest, and the more stringent risk-based performance standards that apply. Lower-risk-tier facilities may not need pre-authorization inspections both because of their comparative simplicity and because inspectors may develop best practices through the pre-authorization inspections of higher-tiered facilities.

In contrast, some policymakers have questioned whether the low inspection rate is due to constraints in the number of chemical facility security inspectors hired by DHS or the availability of appropriated funding. The CFATS regulation states that DHS will inspect the implementation of site security plans at all facilities and requires that facilities resubmit their site security plan every two years for Tier 1 and Tier 2 facilities or three years for Tier 3 and Tier 4 facilities.[32] This requires DHS to perform approximately 1,700 inspections annually in order to inspect each facility's implementation of its site security plan. The DHS has asserted that inspections require two or more inspectors and approximately one week to perform.[33]

The DHS appears to have requested sufficient inspectors to manage the workload associated with a reinspection cycle of every two years for top tier facilities and every three years for lower-tier facilities, but such a staffing level may be insufficient to address the large number of initial regulatory submissions.[34] This level of staffing would appear to require approximately a full cycle of inspections to reduce the backlog created from the initial site security plan submissions. If DHS were to hire additional inspectors, it might reduce the backlog of site security plans but alsorun the risk of having additional unnecessary staff in future years. The DHS might hire temporary or short-term staff to augment the inspector cadre, but the need to train such employees for CFATS-specific inspections may pose challenges.

Finally, because DHS has focused on inspecting those facilities in the highest risk tier, it potentially faces the most complicated inspection environments. Inspections of lower risk tier facilities may pose fewer complications, take less time, and involve fewer inspectors. If so, DHS might quickly and substantially increase the number of facilities inspected by focusing efforts on lower-tier facilities. Through this approach, DHS might gain insight and experience among the inspector cadre while reducing some national risk.[35]

Federal Preemption of State Activities

The original statute did not expressly address the issue of federal preemption of state and local chemical facility security statute or regulation. When DHS issued regulations establishing the CFATS program, DHS asserted that the CFATS regulations would preempt state and local chemical facility security statute or regulation that conflicted with, hindered, posed an obstacle, or frustrated the purposes of the federal regulation.[36] Subsequent to the release of the regulation, Congress amended DHS's statutory authority to state that only in the case of an "actual conflict" would the federal regulation preempt state authority.[37] Few states have established independent chemical facility security regulatory programs, and conflict between the federal and state activities has not yet occurred.[38] The DHS did not identify any state programs that conflict with the CFATS regulations.[39] The DHS has also not altered its regulatory language in response to the statutory amendment.

Advocates for federal preemption call for a uniform security framework across the nation. They assert that a "patchwork" of regulations might develop if states independently develop additional chemical facility security regulations.[40] Variances in security requirements might lead to differing regulatory compliance costs, and companies might suffer competitive disadvantage based on their geographic location.

Supporters of state rights to regulate chemical facility security claim that the federal regulation should be a minimum standard with which all regulated entities must comply. They assert that DHS should allow states to develop more stringent regulations than the federal regulations. Theyclaim such regulations would increase security. Some supporters of state regulation suggest that more stringent, conflicting state regulations should preempt the federal regulations.[41] Such a case might occur if a state regulation mandated the use of a particular security approach at chemical facilities, conflicting with the federal regulation that adopts a performance-based, rather than prescriptive, approach. The desire to retain industries that might relocate faced with increased regulation arguably would temper state inclinations to require overly stringent or incompatible regulations.

Some policymakers may assert that chemical facility security should be left to the states rather than be implemented as a federal regulation. If Congress allows the statutory authority to expire and does not appropriate funds for the further implementation of CFATS, the authority would lapse and states would again be responsible for regulating chemical facility security.

Transparency of Process

The CFATS process involves determining chemical facility vulnerabilities and developing security plans to address them. Information developed in this process is not to be widely or openly disseminated. The CFATS program categorizes this information as CVI and provides penalties for its disclosure. Some advocates have argued for greater transparency in the CFATS process, even if the program does not provide detailed information regarding potential vulnerabilities and specific security measures. They assert that those individuals living in surrounding communities require such non-detailed information to plan effectively and make choices in an emergency.[42]

The current statute and regulation prohibit public disclosure of security-related information. Only specific "covered persons" may access CVI. While acknowledging a legitimate homeland security need to limit dissemination of security information, some policymakers have questioned whether such limitations hinder other efforts. For example, first responders and community representatives have highlighted how such information protection regimes may impede emergency response and the ability of those in the surrounding community to react to emergency situations at the chemical facility.[43] Additionally, worker representatives have raised concerns that these limitations and the lack of mandated inclusion of worker representatives may impede worker input into security plans.[44]

The current information protection regimes for chemical facility security information, CVI under CFATS and Sensitive Security Information (SSI) under the Maritime Transportation Security Act (MTSA), do not contain penalties for incorrectly marking information as protected. Only disclosure of correctly marked information is penalized. Additionally, the chemical facility is responsible for identifying and appropriately marking protected information. These information markings only would be assessed in the case of dispute. As was asserted during congressional oversight, this disparity may lead to a tendency by regulated entities, in order to protect themselves against potential liability or scrutiny, to erroneously limit dissemination of information that should be made available to the public.[45]

Congressional investigation indicated that documents related to a 2007 explosion at a Bayer CropScience chemical facility in West Virginia were incorrectly labeled as protected from disclosure.[46] The DHS regulated this chemical facility under MTSA, not CFATS.[47] In this case, security information was protected from disclosure as SSI, an information protection regime similar to CVI. Company officials broadly applied SSI markings to facility documents partly in hopes of avoiding a public debate on the use and storage of particular chemicals at the facility.[48] This revelation led to questions regarding the application and oversight of such protective markings.

Additionally, the existing statute contains no provisions explicitly protecting or allowing for concerned covered persons to divulge CVI or to challenge the categorization of information as protected in an attempt to inform authorities about security vulnerabilities or other weaknesses. Depending on the circumstances, those individuals might be penalized for their disclosure of protected information. The CFATS regulations, reflecting this inherent tension, provide for a point of contact to which such information might be revealed, but also state “Section 550 did not give DHS authority to provide whistleblower protection, and so DHS has not incorporated specific whistleblower protections into this regulation.”[49]

Definition of Chemical Facility

The DHS regulates both entities that possess and entities that manufacture chemicals of interest. Thus, the term chemical facility encompasses many types of facilities, including agricultural facilities, universities, and others. With DHS defining chemical facilities according to possessionof a substance of concern, facilities not part of the chemical manufacturing and distributing chain have become regulated facilities. Stakeholders have expressed concern that the number of entities so regulated might be unwieldy and that the regulatory program might focus on many chemical facilities that pose little risk rather than on those facilities that pose more substantial risk. For example, during the rulemaking process, DHS received commentary and revised its regulatory threshold for possession of propane, stating:

> DHS, however, set the [screening threshold quantities] for propane in this final rule at 60,000 pounds. Sixty thousand pounds is the estimated maximum amount of propane that nonindustrial propane customers, such as restaurants and farmers, typically use. The Department believes that non-industrial users, especially those in rural areas, do not have the potential to create a significant risk to human life or health as would industrial users. The Department has elected, at this time, to focus efforts on large commercial propane establishments but may, after providing the public with an opportunity for notice and comment, extend its [CFATS] screening efforts to smaller facilities in the future. This higher [screening threshold quantity] will focus DHS's security screening effort on industrial and major consumers, regional suppliers, bulk retail, and storage sites and away from non-industrial propane customers.[50]

Similarly, academic institutions have asserted that DHS should not apply CFATS regulations to them because of the dispersed nature of chemical holdings at colleges and universities. These institutions claim that regulatory compliance costs would not be commensurate with the risk reduction.[51] While the regulatory compliance costs likely decrease at lower-risk tiers compared to higher-risk tiers, all regulated entities bear compliance costs as continued annual expenses.

As mentioned above, the statutory authority underlying CFATS exempts several types of facilities, including water and wastewater treatment facilities. The federal government does not regulate water and wastewater treatment facilities for chemical security purposes. Instead, current chemical security efforts at water and wastewater treatment facilities are voluntary in nature.[52] Some advocacy groups have called for inclusion of currently exempt facilities, such as water and wastewater treatment facilities.[53] Some drinking water and wastewater treatment facilities possess large amounts of potentially hazardous chemicals, such as chlorine, for purposes such as disinfection.[54] Advocates for their inclusion in security regulations cite the presence of such potentially hazardous chemicals and their relative proximity to population centers as reasons to mandate security measures for such facilities. In contrast, representatives of the water sector point to the critical role that water and wastewater treatment facilities play in daily life. They caution against including these facilities in the existing regulatory framework because of the potential for undue public impacts. They cite, for example, loss of basic fire protection and sanitation services if the federal government orders a water or wastewater utility to cease operations for security reasons or failure to comply with regulation.[55]

If Congress were to remove the drinking water and wastewater treatment facility exemption, the number of regulated facilities might substantially increase, placing additional burdens on the CFATS program. The United States contains approximately 52,000 community water systems and 16,500 wastewater treatment facilities.[56] These facilities vary substantially in size and service. The number of regulated facilities would depend on the criteria used to determine inclusion, such as chemical possession or number of individuals served. It is likely that only a subset of these facilities would meet a regulatory threshold.[57] A DHS official testified that approximately 6,000 facilities would likely meet the CFATS threshold.[58]

Inherently Safer Technologies

Previous debate on chemical facility security has included whether to mandate the adoption or consideration of changes in chemical processes to reduce the potential consequences following a successful attack on a chemical facility. Suggestions for such changes have included reducing the amount of chemical stored onsite and changing the chemicals used. In previous congressional debate, these approaches have been referred to as inherently safer technologies or methods to reduce the consequences of a terrorist attack.

A fundamental challenge for inherently safer technologies is how to compare one technology with its potential replacement. It is challenging to unequivocally state that one technology is inherently safer than the other without adequate metrics. Risk factors may exist outside of the comparison framework.[59] Some experts have asserted that the metrics for comparing industrial processes are not yet fully established and need additional research and study.[60] The National Academies have recommended that DHS support research and development to foster cost-effective, inherently safer chemistries and chemical processes.[61] A facility might consider many additional factors beyond homeland security implications when weighing the applicability and benefit of switching from one process to another. These factors include cost, technical challenges regarding implementation in specific situations, supply chain impacts, quality and availability of end products, and indirect effects on workers.[62]

Supporters of adopting these approaches as a way to improve chemical facility security argue that reducing or removing these chemicals from a facility will reduce the incentive to attack the facility. They suggest that reducing the consequences of a release also lowers the threat from terrorist attack and mitigates the risk to the surrounding populace. They point to facilities that have voluntarily changed amounts of chemicals on hand or chemical processes in use as examples that facilities can implement such an approach in a cost-effective, practical fashion.[63]

Opponents of mandating what proponents call inherently safer technologies question the validity of the approach as a security tool and the government's ability to effectively oversee its implementation. Industrial entities assert that process safety engineers within the regulated industry already employ such approaches and that these are safety, not security, methods. They assert that process safety experts and business executives should determine the applicability and financial practicality of changing existing processes at specific chemical facilities.[64] One industry survey stated that, of those respondents that assessed using alternative chemicals or processes, 66.4% determined such alternatives were not technically feasible.[65] Opponents of an inherently safer technology mandate also state concern that few existing alternative approaches are well understood with regard to their unanticipated side effects. They claim that researchers should continue to study these alternative approaches rather than immediately apply them, since unanticipated side effects could injure business and other interests.[66] A third opposing view questions whether the federal government contains the required technical expertise to adjudicate the practicality and benefit of alternative technological approaches. Holders of this view raise concerns that the federal government may not possess the required knowledge or expertise to judge whether a particular site can implement alternative technology, even if the alternative theoretically provides benefits over existing technology.[67]

Some industry representatives have asserted that an inherently safer technology mandate might have a potentially significant negative financial impact.[68] Regulated entities incur a cost when meeting existing CFATS requirements, and small businesses may be challenged to make necessary capital investments. In its interim final rule, DHS estimated that CFATS "may have a significant economic impact on a substantial number of small entities."[69] Because of the performance-based nature of the regulatory requirement, it is difficult to detail the exact impact on small businesses.70

Adding an inherently safer technology requirement might increase the cost of CFATS compliance and might disproportionately affect small entities not already incorporating such activities in their business processes. Policymakers in previous Congresses highlighted the issue of small business impact, especially in the context of requiring additional measures that might hurt productivity.

POLICY OPTIONS

The statutory authority for CFATS expires on November 18, 2011. The 112th Congress may address chemical facility security through several options. Congress might extend further the existing statutory authority by revising or repealing its sunset provision; codify the existing regulations; amend the existing statutory authority; address existing programmatic activities; or restrict or expand the scope of chemical facility security regulation.

If Congress does not act and allows the statutory authority to expire, the authority for the application and enforcement of the CFATS regulations may be questioned. In the case where Congress allows the statutory authority to expire, but Congress appropriates funds for enforcing the CFATS program, DHS will likely be able to enforce the CFATS regulations. The Government Accountability Office (GAO) has found that in the case where a program's statutory authority expires, but Congress explicitly appropriates funding for it, the program may continue to operate without interruption.71 If Congress allows the statutory authority to expire and also does not appropriate funding for implementing the CFATS program, the CFATS regulations will likely also lapse. In this case, the states would likely become the primary source of any chemical facility security regulation.

Maintain the Existing Regulatory Framework

The existing statutory authority places much of the CFATS regulatory framework at the discretion of the Secretary of Homeland Security. The DHS is still in the process of implementing these regulations and has not yet determined their efficacy. Congressional oversight of their implementation, enforcement, and efficacy may play a key role in determining the sufficiency of the existing authority and regulations. Congress might choose to maintain the existing regulations by extending the statutory authority's sunset date or codifying the existing regulations. Also, as noted above, allowing the statutory authority to expire could in effect maintain the existing regulatory framework if Congress continues to fund implementation, although this might lead to litigation.

Extend the Sunset Date

Congressional policymakers might choose to extend the current statutory authority for a fixed or indefinite time. Congress has enacted a series of limited extensions of the statutory authority since its inception. The Continuing Appropriations Act, 2012 (P.L. 112-36) extends the statutory authority through November 18, 2011. The Obama Administration requested for FY2012 an extension of the statutory authority to October 4, 2013.[72] Extending the existing statutory authority may provide regulated entities continuity and protect them from losing those resources already expended in regulatory compliance. An extension may allow assessment of the efficacy of the existing regulations and inclusion of this information in any future attempts to revise or extend DHS's statutory authority. Moreover, since DHS is in the process of implementing current regulations, some policymakers argue for a simple extension without changing statutory requirements.

In addition to requesting a two-year extension of the statutory authority, the Obama Administration also supports enacting a permanent statutory authority.[73] Congress might make the existing program permanent by removing the sunset date entirely. Some chemical manufacturers support converting the existing program into a permanent program.[74] The removal of the sunset date would maintain the current discretion granted to the Secretary of Homeland Security to develop regulations and might allow assessment of the efficacy of the existing regulations. Making the existing statute permanent would provide consistency in authority and remove the statutory pressure to reauthorize the program. The presence of a sunset date for the statutory authority increases the likelihood of congressional attention to chemical facility security as a legislative topic. Some advocates who wish for more regular congressional review of the statute might oppose removing the sunset date.

Codify the Existing Regulations

Congressional policymakers might choose to affirm the existing regulations by codifying them or their principles in statute. Such codification could reduce the discretion of the Secretary of Homeland Security to alter the CFATS regulations in the future. The existing statutory authority grants broad discretion to the Secretary to develop many elements of the CFATS regulations. Future Secretaries may choose to alter its structure or approach and still comply with the existing statute. Policymakers might identify specific components of the existing regulation that they wish any future regulation to retain and codify those portions. Specifying these components might limit the ability of the Secretary to react to changing circumstance, gained experience, and new knowledge. On the other hand, the codified portions might enhance the regulated community's ability to plan for future expenses and requirements.

Alter the Existing Statutory Authority

Congressional policymakers might choose to alter the existing statutory authority to modify the existing regulations, address stakeholder concerns, or broadly change the regulatory program.

Accelerate or Decelerate Compliance Activities

The DHS bases its schedule for facility CFATS compliance on the chemical facility's assigned risk tier. Those chemical facilities assigned to higher risk tiers have a more accelerated compliance and resubmission schedule than those assigned to lower risk tiers. Congressional policymakers might attempt to accelerate the compliance schedule by increasing funding available to DHS for CFATS, thereby increasing the ability of DHS to provide feedback to regulated entities, review submissions, and inspect facilities filing site security plans. Additional funding might reduce or mitigate inefficiencies or delays related to DHS processing of submissions.

Alternatively, policymakers might provide DHS with the authority to use third parties as CFATS inspectors. The DHS could then augment the number of CFATS inspectors to meet increased demand or delegate inspection authority to state and local governments. Third-party inspectors might allow DHS to draw on expertise outside of the federal government in assessing the efficacy of the implemented site security activities. The DHS may need to define the roles and responsibilities of these inspectors and how DHS will assess and accredit their qualifications. The DHS has stated its intent to issue a rulemaking regarding the use of third-party inspectors but has not yet done so.[75] The use of third-party inspectors might lead to concerns about equal treatment of chemical facilities by different third-party inspectors, and questions about whether homeland security inspections of this type are an inherently governmental responsibility that only federal employees should perform.

Finally, Congress might determine that DHS has sufficient resources to accelerate compliance activities but is restrained by some other procedural factor. Congressional policymakers might direct DHS to streamline its review process, reduce the timeframe for response and interaction with regulated entities, or otherwise enact process improvements.

Congressional policymakers might choose to slow the implementation schedule of the chemical facility security regulations. Concern about the impact of the regulation on small businesses or other entities might lead to a decelerated compliance schedule. The DHS has already implemented select regulatory extensions for certain agricultural operations.[76] Congressional policymakers might direct DHS to provide longer submission, implementation, and resubmission timelines for those regulated entities that might suffer disproportionate economic burdens from compliance.

Incorporate Additional Facility Types

Policymakers might remove some or all of the statutory exclusions from the CFATS program. The DHS and the Environmental Protection Agency (EPA) have called for additional authorities to regulate water and wastewater treatment facilities:

> The Department of Homeland Security and the Environmental Protection Agency believe that there is an important gap in the framework for regulating the security of chemicals at water and wastewater treatment facilities in the United States. The authority for regulating the chemical industry purposefully excludes from its coverage water and wastewater treatment facilities. We need to work with the Congress to close this gap in the chemical security authorities in order to secure chemicals of interest at these facilities and protect the communities they serve. Water and wastewater treatment facilities that are determined to be high-risk due to the presence of chemicals of interest should be regulated for security in a manner that is consistent with the CFATS risk and performance-based framework while also recognizing the unique public health and environmental requirements and responsibilities of such facilities.[77]

The EPA has testified that the Obama Administration believes that EPA should be the lead agency for chemical security for both drinking water and wastewater systems, with DHS supporting EPA's efforts.[78] The EPA also supports providing states with an important role in regulating chemical security at water systems, including determinations, auditing, and inspecting.[79]

In addition, DHS supports modifying the existing exemption for (1) MTSA facilities to increase security at these facilities to the CFATS standard and (2) facilities regulated by the Nuclear Regulatory Commission to clarify the scope of the exemption.[80]

If Congress provides the executive branch with statutory authority to regulate water and wastewater treatment facilities for chemical security purposes, it may weigh several policy decisions. Among these choices are which facilities should be regulated; how stringent such security measures should be; what federal agency should oversee them; and whether compliance with these security measures is practicable given the public nature of many water and wastewater treatment facilities.

One option for congressional policymakers might be to include water and wastewater treatment facilities under the existing CFATS regulations, effectively removing the exemption currently in statute. This would place water and wastewater treatment facilities on par with other possessors of chemicals of interest. The DHS would provide oversight of all regulated chemical facilities.[81] Opponents might claim that activities under CFATS, such as vulnerability assessment, duplicate existing requirements under the Safe Drinking Water Act.[82] Also, opponents of such an approach cite the essential role that water and wastewater treatment facilities play in daily life and assert that several authorities available to DHS under CFATS, such as the ability to require a facility to cease operations, are inappropriate if applied to a municipal utility.[83]

Another option might be to grant statutory authority to regulate water and wastewater treatment facilities for security purposes to EPA or require DHS to consult with EPA regarding its regulation of water and wastewater treatment facilities. Since water treatment facilities must provide a vulnerability assessment to EPA, some facilities might view regulation under CFATS as redundant in this context. Some industry representatives have expressed concern regarding the effects of multiple agencies regulating security at drinking water and wastewater treatment facilities.[84] They assert that municipalities that operate both types of facilities might face conflicting regulations and guidance if different agencies regulate drinking water and wastewater treatment facilities. These stakeholders suggest that EPA retaining the lead for water and wastewater facilities would be more efficient. Following prior debate on chemical facility security, Congress provided statutory authority for chemical security to DHS. This separated DHS security responsibilities from the public health and safety responsibilities given to EPA. Providing one agency the authority to oversee safety and security operations may reduce the potential for redundancy and other inefficiencies but also might increase stakeholder reluctance to voluntarily consult on security issues.[85]

If policymakers assign responsibility for chemical facility security at different facilities to different agencies, each agency affected will promulgate separate rules. These rules may be similar or different depending on the agencies' statutory authority, interpretation of that authority, and ability of the regulated entities to comply as well as any interagency coordination that might occur. Congress may wish to assess the areas where such facilities are similar and different in order to provide authorities that meet any unique characteristics.

Any new regulation of drinking water and wastewater treatment facilities is likely to cause the regulated entities, and potentially the federal government, to incur some costs. Representatives of the water and wastewater sectors argue that local ratepayers will eventually bear the capital and ongoing costs incurred due to increased security measures.[86] Congressional policymakers may wish to consider whether the regulated entities should bear these costs, as is done for other regulated chemical facilities, and by those ratepayers they serve or by the taxpayers in general through financial assistance to the regulated entities. Additionally, if inclusion of other facility types significantly increases the number of regulated entities, DHS may require additional funds to process regulatory submissions and perform required inspections.

Harmonize Regulations

Other security provisions, such as MTSA, apply to some facilities exempt from the existing chemical facility security regulations. The DHS supports modifying the existing exemption for MTSA facilities to increase security at these facilities to the CFATS standard and modifying the existing exemption for facilities regulated by the Nuclear Regulatory Commission to clarify the scope of the exemption.[87] The EPA has testified that the Obama Administration believes that DHS should be responsible for ensuring consistency of high-risk chemical facility security across all critical infrastructure sectors.[88] If Congress modifies these exemptions, conflicts may arise between requirements under chemical facility security regulations and these other provisions. One approach to resolving these conflicts is to identify which statute would supersede the others, providing a single statutory requirement. Critics of such an approach might assert that the superseding statute does not contain all of the protections present in the other statutes. Another approach might be to require agencies to generally harmonize the regulations implementing each statute. Regulatory agencies might identify and determine the best ways to meet statutory requirements while also limiting regulatory duplication or contradiction. Such harmonization might reduce the regulatory burden on companies possessing facilities regulated under two frameworks, such as MTSA and CFATS, by allowing a single security approach to the regulations. The DHS has established a joint NPPD/U.S. Coast Guard working group to evaluate and, where appropriate, implement methods to harmonize the CFATS and MTSA regulations.[89] In contrast, if the process of harmonization leads to a significant increase in security requirements, the regulatory burden faced by industry might increase.

Consider Inherently Safer Technologies

Congressional policymakers may choose to address the issue of inherently safer technologies, sometimes called methods to reduce the consequences of terrorist attack. One approach might be to mandate the implementation of inherently safer technologies for a set of processes. Another might be to mandate the consideration of implementation of inherently safer technologies with certain criteria controlling whether implementation is required. A third approach might be to mandate the development of a federal repository of inherently safer technology approaches and consideration of chemical processes against those options listed in the repository. Stakeholders might assess and review the viability of applying these inherently safer approaches at lower cost if such information were centralized and freely available. Alternatively, policymakers might establish an incentive-based structure outside of the chemical facility security mandate to encourage the adoption of inherently safer technologies by regulated entities. Lastly, congressional policymakers might choose to not require any consideration or adoption of inherently safer technology approaches.

The Obama Administration has given some support to the use of inherently safer technologies to enhance security at high-risk chemical facilities. It has established a series of principles directing its policy:

- The Administration supports consistency of inherently safer technology approaches for facilities regardless of sector.
- The Administration believes that all high-risk chemical facilities, Tiers 1-4, should assess [inherently safer technology] methods and

report the assessment in the facilities' site security plans. Further, the appropriate regulatory entity should have the authority to require facilities posing the highest degree of risk (Tiers 1 and 2) to implement inherently safer technology methods if such methods demonstrably enhance overall security, are determined to be feasible, and, in the case of water sector facilities, consider public health and environmental requirements.

- For Tier 3 and 4 facilities, the appropriate regulatory entity should review the inherently safer technology assessment contained in the site security plan. The entity should be authorized to provide recommendations on implementing inherently safer technologies, but it would not have the authority to require facilities to implement the inherently safer technology methods.
- The Administration believes that flexibility and staggered implementation would be required in implementing this new inherently safer technology policy.[90]

A congressional mandate for regulated entities to adopt or consider adopting inherently safer technologies may lead regulated entities to consider factors such as homeland security impact in their chemical process assessments. Some experts assert that existing chemical process safety activities consider and assess inherently safer technology approaches though not necessarily in a homeland security context.[91] These assessments may lead to changes in chemical process when deemed safer, more reliable, and cost-effective. The extent to which homeland security impact has factored into these industry decisions is unknown, but DHS has identified cases where chemical facilities have voluntarily modified chemical processes to lower their CFATS tier. An additional complication to assessing inherently safer technology is the varying amounts and quality of information available regarding industrial implementation of inherently safer technologies. While some facilities have converted to processes generally deemed as inherently safer, other facilities may not have sufficient information available to effectively assess the impacts from changing existing processes to ones considered inherently safer.[92] The differences that exist among chemical facilities, in terms of chemical process, facility layout, and ability to finance implementation, may challenge mandatory implementation of inherently safer technologies at regulated entities. Even the mandatory consideration of inherently safer technologies may place a financial burden on some small regulated entities. Congress might limit mandatory measures to those facilities considered by DHS to pose the most risk or might provide such financial assistance to regulated facilities.[93]

Policymakers might choose to try to further incentivize regulated entities to adopt inherently safer technologies. Under the CFATS regulations, facilities that adopt inherently safer technologies might change their assigned risk tier by reducing the amount of chemicals of interest on hand. Policymakers might provide regulated entities that adopt inherently safer technologies with financial or regulatory incentives. Alternatively, policymakers might direct DHS or another agency to perform inherently safer technology assessments for regulated entities, transferring the cost of such assessment from the facility to the federal government.[94] The regulated entity or the overseeing agency might use the results of these assessments to guide implementation.

Modify Information Security Provisions

Congressional policymakers might choose to increase transparency in the CFATS process by altering the information security provisions of the program. Such an approach might include increasing the number and type of individuals granted access to CVI, improving information exchange with first responders, and adjusting the manner by which courts and administrative proceedings handle CVI. The Obama Administration has testified that CVI is a distinct information protection regime and expressed support for maintaining CVI in its current form.[95]

Congress might choose to amend the existing statutory authority to address policy concerns. For example, while still maintaining disclosure prohibitions for vulnerability or security related information, Congressional policymakers might require that DHS gather and document comments and information. Such input might come from outside groups, worker organizations, or other trade representatives through formal and informal mechanisms or by the solicitation, development, and use of industry best practices. Policymakers might direct DHS to make specific types of information, such as the results of enforcement activities or the approval of successful implementation of a site security plan, more generally available. By mandating the inclusion of such information gathering or the release of specific information, congressional policymakers might facilitate greater cooperation between various stakeholder groups. Conversely, such requirements may raise concerns about the degree of security given to the protected information, since more individuals will participate in its development and analysis, perhaps increasing the ability of malicious persons to use such information for targeting purposes. As more information about the vulnerability assessment process and the results of the security process becomes available, the potential that adversaries might combine this disparate information to obtain insight into a security weakness could result in a dangerous synergy of information. Congressional policymakers might require that the executive branch or another entity identify the threats or vulnerabilities that might accrue from release of a greater amount of chemical facility security information prior to implementing such a policy change.[96]

Congressional policymakers can choose as a policy option to alter the information protection regime afforded to chemical facility security information by specifically expanding access to first responders. The existing regulation explicitly states that information developed in response to other laws or regulations, such as Emergency Planning and Community Right-to-Know Act, are not protected from disclosure. Enhancing first responder access to such information might minimize perceived barriers to disclosing information during an accident. For example, Congress might mandate that each jurisdiction containing a regulated chemical facility contain a first responder designated as a covered individual.

Congressional policymakers also can choose to further limit dissemination of CVI so as to increase barriers to its release if that is a policy goal. Congress might prohibit DHS from sharing such information outside of the federal government or set particular criteria that would allow CVI access to state and local officials. Limiting the number of individuals with access to CVI may make it more difficult for those wishing to do harm to obtain technical or operational security information. Conversely, state and local officials may not support such an approach, as limitations on distribution may also adversely affect emergency response at a regulated facility or inhibit the ability of state and local law enforcement officials to provide targeted protection of particular chemical facility assets.

Policymakers might also choose to address the issue of identifying and marking protected information by mandating review of marked documents. Congressional policymakers might place this responsibility to review and certify marked information on the chemical facility. Alternatively, the federal government might review and certify documents marked CVI on a regular basis. Industry representatives may not support such a review requirement due to the additional regulatory burden caused by the review. Additionally, while such review might potentially limit incorrect marking, it may inhibit information reporting by regulated entities to the federal government. Additionally, absent a penalty for incorrect marking, it is unclear how to ensure compliance.

Congressional policymakers may also address concerns raised regarding the ability of concerned individuals to report misdeeds by creating a "whistleblower" reporting mechanism.[97] One approach might be to codify the current mechanism of reporting such concerns specific to DHS or a similar federal entity, such as an agency Inspector General. Alternatively, Congress can create a more general exemption to the penalties arising from disclosure of protected information for those individuals who report such concerns to federal officials if that is needed to protect whistleblowers. As part of a whistleblower mechanism, policymakers might choose to extend protections against retaliation or other job-related actions to those individuals availing themselves of current or newly established reporting mechanisms.

Preempt State Regulations

The 110th Congress addressed the issue of federal preemption of state chemical facility security statutes and regulations by placing in statute the requirement that only when an "actual conflict" occurs between state and federal regulation will the state regulation be preempted.[98] Congressional policymakers may choose to further limit the cases where federal regulation would preempt state regulation by affirming the right of states to make chemical facility security regulations that are more stringent than federal regulation even if they conflict. Alternatively, policymakers may choose to increase the number of cases where federal regulations preempt those of a state by expanding the types of conflict, beyond "actual," that will lead to preemption.

CONGRESSIONAL ACTION

The annual appropriations process provides funding for implementation of chemical facility security regulation. The Continuing Appropriations Act, 2012 (P.L. 112-36) extends the statutory authority through November 18, 2011. While P.L. 112-10 provides a FY2011 expenditure level for the National Protection and Programs Directorate—Infrastructure Protection and Information Security appropriations account, it does not specify a level for chemical facility security regulation.

Extend the Existing Authority

The current statutory authority expires on November 18, 2011. Congress is considering extending the existing authority through appropriations and authorization legislation. The Obama Administration has requested an extension of the existing statutory authority in each budget request. It requested an extension of the existing statutory authority to October 4, 2011, in the FY2011 budget and an extension to October 4, 2013, in the FY2012 budget.[99]

H.R. 901

H.R. 901, the Chemical Facility Anti-Terrorism Security Authorization Act of 2011, was ordered reported as amended by the House Committee on Homeland Security. The act would amend the Homeland Security Act of 2002 with provisions authorizing DHS oversight of chemical facility security. The provisions of H.R. 901 generally match the existing statutory authority. H.R. 901 would also authorize appropriation of approximately $90 million annually from FY2012 through FY2018. The statutory authority would expire on September 30, 2018. In addition, the DHS would be required to approve or disapprove of vulnerability assessments and site security plans within 180 days of receipt and provide technical support to regulated entities qualifying as small businesses. The DHS would issue guidance on how alternative background checks would meet in full or in part any background check personnel security requirement. Finally, the DHS would be required to report to select congressional committees regarding its success at meeting the 180 day requirement, efforts to harmonize CFATS and MTSA regulations, and on the number of jobs created or eliminated due to CFATS regulation.

H.R. 901 was also referred to the House Committee on Energy and Commerce to the Subcommittee on Environment and the Economy. The subcommittee has taken no further action on this bill.

H.R. 908

H.R. 908, the Full Implementation of the Chemical Facility Anti-Terrorism Standards Act, was ordered reported as amended by the House Committee on Energy and Commerce. The act would extend the existing statutory authority to October 4, 2018. H.R. 908 would authorize appropriations of $89.92 million for each fiscal year from FY2012 through FY2018. It would allow the Secretary of Homeland Security to accept security background checks conducted for other purposes. Finally, it would also allow holders of Transportation Worker Identification Credential cards access to CFATS-regulated facilities.

H.R. 916

H.R. 916, the Continuing Chemical Facilities Antiterrorism Security Act of 2011, was referred to the House Committee on Energy and Commerce and the House Committee on Homeland Security. The act would extend the existing statutory authority to October 4, 2015. It would also amend the Homeland Security Act of 2002 to direct the Secretary of Homeland Security to establish a voluntary chemical security training program and a voluntary chemical security exercise program. Finally, it would authorize such sums as necessary for these programs.

H.R. 2017/P.L. 112-33

H.R. 2017/P.L. 112-33, the Continuing Appropriations Act, 2012, became law on September 30, 2011. It extended the existing statutory authority to October 4, 2011.

H.R. 2608/P.L. 112-36

H.R. 2608/P.L. 112-36, the Continuing Appropriations Act, 2012, became law on October 5, 2011. It extends the existing statutory authority to November 18, 2011.

S. 473

S. 473, the Continuing Chemical Facilities Antiterrorism Security Act of 2011, was ordered reported with an amendment by the Senate Committee on Homeland Security and Governmental Affairs. S. 473 would extend the existing statutory authority to October 4, 2014. In addition, it would amend the Homeland Security Act of 2002 to direct the Secretary of Homeland Security to establish a voluntary chemical security training program, a voluntary chemical security exercise program, a voluntary technical assistance program, and a chemical facility security advisory board. S. 473 would authorize such sums as necessary for the programs and board.

Modify the Existing Authority

Legislation has been introduced in both chambers that would modify the existing authority.

H.R. 225

H.R. 225, the Chemical Facility Security Improvement Act of 2011, was referred to the House Committee on Energy and Commerce and the House Committee on Homeland Security. The act would prohibit the Secretary of Homeland Security from approving a chemical facility site security plan if the plan did not meet or exceed existing state or local security requirements. It would allow the Secretary of Homeland Security to mandate the use of specific security measures in site security plans. The bill would also cause CVI to be treated as sensitive security information in both general and legal proceedings. Finally, the act would no longer prohibit thirdparty individuals from bringing suit in court to require the Secretary of Homeland Security to enforce chemical facility security regulations against a chemical facility.

S. 709

S. 709, the Secure Chemical Facilities Act, was referred to the Senate Committee on Homeland Security and Governmental Affairs. The act would codify aspects of the CFATS regulation. It would require facilities to evaluate whether the facility could reduce the consequences of an attack by using a safer chemical or process. The act would authorize DHS to require implementation of those safer measures if a facility has been classified as one of the highest-risk facilities, implementation of safer measures is feasible, and implementation would not increase risk overall by shifting risk to another location. Among other provisions, S. 709 also would increase the participation of employees and employee representatives in developing security plans. S. 709 would alter the current information control regime, aligning it with that for sensitive security information. Finally, S. 709 would allow citizens to file suit against the Secretary of Homeland Security or submit a petition to the Secretary to enforce compliance with statute.

H.R. 2890

H.R. 2890, a bill to expand homeland security at public water systems and treatment works by allowing the Secretary of Homeland Security to include these facilities in the Chemical Facility Anti-Terrorism Standards program, was referred to the House Committee on Energy and Commerce and the House Committee on Transportation and Infrastructure. The act would expand chemical facility security regulation to public water systems and wastewater treatment facilities and direct the President to delegate such authority from the Secretary of Homeland Security to the EPA Administrator.

S. 711

S. 711, the Secure Water Facilities Act, was referred to the Senate Committee on Environment and Public Works. The act would authorize the EPA Administrator to regulate community water systems and wastewater treatment facilities for security purposes. S. 711 also would authorize implementation of methods to reduce the consequences of a chemical release from an intentional act. Among other provisions, the Administrator would be directed to promulgate regulations as necessary to prohibit the unauthorized disclosure of controlled information. S. 711 would authorize the Administrator to provide grants or enter into cooperative agreements with states or regulated entities to assist in regulatory compliance.

End Notes

[1] For example, New Jersey, Maryland, and New York each enacted laws addressing security at chemical facilities.

[2] Section 550, P.L. 109-295, Department of Homeland Security Appropriations Act, 2007.

[3] The original statute expired on October 4, 2009, three years after enactment. The Department of Homeland Security Appropriations Act, 2010 (P.L. 111-83) extended the existing statutory authority an additional year. The Continuing Appropriations Act, 2011 (P.L. 111-242) extended the statutory authority through December 3, 2010. The second continuing resolution (P.L. 111-290) extended the statutory authority through December 18, 2010. The third continuing resolution (P.L. 111-317) extended the statutory authority through December 21, 2010. The Continuing Appropriations and Surface Transportation Extensions Act, 2011 (P.L. 111-322) extended the statutory authority through March 4, 2011. The Further Continuing Appropriations Amendments, 2011 (P.L. 112-4) extended the statutory authority through March 18, 2011. The Additional Continuing Appropriations Amendments, 2011 (P.L. 112-6) extended the statutory authority through April 8, 2011. The Further Additional Continuing Appropriations Amendments, 2011 (P.L. 112-8) extended the statutory authority through April 15, 2011. The Department of Defense and Full-Year Continuing Appropriations Act, 2011 (P.L. 112-10) and the Continuing Appropriations Act, 2012 (P.L. 112-33) both extended the statutory authority through October 4, 2011. The Continuing Appropriations Act, 2012 (P.L. 112-36) extends the statutory authority through November 18, 2011.

[4] 72 *Federal* Register 17688-17745 (April 9, 2007). An interim final rule is a rule that meets the requirements for a final rule and that has the same force and effect as a final rule, but that contains an invitation for further public comment on its provisions. After reviewing comments to the interim final rule, an agency may modify the interim final rule and issue a "final" final rule. The DHS first issued a proposed rule in December 2006 and solicited public comments. 71 *Federal Register* 78276-78332 (December 28, 2006).

[5] 72 *Federal Register* 65396-65435 (November 20, 2007).

[6] According to the White House Office of Management and Budget, a performance standard is a standard that states requirements in terms of required results with criteria for verifying compliance but without stating the methods for achieving required results. A performance standard may define the functional requirements for the item, operational requirements, and/or interface and interchangeability characteristics. A performance standard may be

viewed in juxtaposition to a prescriptive standard which may specify design requirements, such as materials to be used, how a requirement is to be achieved, or how an item is to be fabricated or constructed.

Office of Management and Budget, The White House, "Federal Participation in the Development and Use of Voluntary Consensus Standards and in Conformity Assessment Activities," *Circular A-119*, February 10, 1998. For example, a performance standard might require that a facility perimeter be secured, while a prescriptive standard might dictate the height and type of fence to be used to secure the perimeter.

[7] High-risk facilities may develop vulnerability assessments and site security plans using alternative security programs so long as they meet the tiered, performance-based requirements of the interim final rule.

[8] P.L. 110-161, the Consolidated Appropriations Act, 2008, Section 534.

[9] The budget request for the Infrastructure Security Compliance Project contains the funding and personnel allocations for implementing the CFATS regulations.

[10] Oral testimony of Rand Beers, Under Secretary, National Protection and Programs Directorate, Department of Homeland Security, before the House Homeland Security Committee, Subcommittee on Cybersecurity, Infrastructure Protection, and Security Technologies, February 11, 2011. As of July 2010, DHS had hired 88 field personnel, including 11 regional commanders. Office of Infrastructure Protection, National Protection and Programs Directorate, Department of Homeland Security, *Update on Implementation of the Chemical Facility Anti-Terrorism Standards and Development of Ammonium Nitrate Regulations-2010 Chemical Sector Coordinating Council Security Summit*, July 7, 2010.

[11] Infrastructure Security Compliance Division, Office of Infrastructure Protection, National Protection and Programs Directorate, Department of Homeland Security, *Chemical Facility Anti-Terrorism Standards*, January 27, 2011.

[12] Department of Homeland Security, "DHS Notifies Chemical Facilities of Revised Tiering Assignments," July 5, 2011, http://www.dhs.gov/files/programs/cfats-revised-tiering-assignments.shtm.

[13] Society of Chemical Manufacturers and Affiliates, "DHS Provides latest on CFATS and Tiering at Chemical Sector Security Summit," http://www.socma.com/tags/printerFriendly.cfm?pageid=3109.

[14] CRS analysis of Department of Homeland Security data from Infrastructure Security Compliance Division, Office of Infrastructure Protection, National Protection and Programs Directorate, Department of Homeland Security, *Chemical Facility Anti-Terrorism Standards*, January 27, 2011; Personal communication with Department of Homeland Security, September 15, 2011; and AcuTech Consulting Group, *A Survey of CFATS Progress in Securing the Chemical Sector*, September 6, 2011.

[15] Department of Homeland Security, *Chemical Facility Anti-Terrorism Standards Interim Final Rule Regulatory Assessment,* DHS-2006-0073, April 1, 2007, p. 15.

[16] In July 2007, DHS provided testimony that formal site inspections of a selected group of facilities would begin by the end of the calendar year (Testimony of Robert B. Stephan, Assistant Secretary for Infrastructure Protection, National Protection and Programs Directorate, Department of Homeland Security, before the House Committee on Homeland Security, Subcommittee on Transportation Security and Infrastructure, July 24, 2007). In December 2007, DHS provided testimony that facility inspection would begin in fall of 2008 (Testimony of Robert B. Stephan, Assistant Secretary for Infrastructure Protection, National Protection and Programs Directorate, Department of Homeland Security, before the House Committee on Homeland Security, Subcommittee on Transportation Security and Infrastructure, December 13, 2007). In 2009, DHS provided testimony that inspections would begin in the first quarter of FY2010 (Testimony of Philip Reitinger, Deputy Under Secretary, National Protection and Programs Directorate, Department of Homeland Security, before the House Committee on Homeland Security, June 16, 2009). The DHS now states that it expects to inspect all Tier 1 facilities by the end of calendar year 2011

(Oral testimony of Rand Beers, Under Secretary, National Protection and Programs Directorate, Department of Homeland Security, before the House Homeland Security Committee, Subcommittee on Cybersecurity, Infrastructure Protection, and Security Technologies, February 11, 2011).

[17] Testimony of Rand Beers, Under Secretary, National Protection and Programs Directorate, Department of Homeland Security, before the Senate Committee on Homeland Security and Governmental Affairs, March 3, 2010.

[18] Oral testimony of Rand Beers, Under Secretary, National Protection and Programs Directorate, Department of Homeland Security, before the Senate Committee on Homeland Security and Governmental Affairs, March 3, 2010.

[19] Personal communication with Department of Homeland Security, September 15, 2011.

[20] Oral testimony of Rand Beers, Under Secretary, National Protection and Programs Directorate, Department of Homeland Security, before the House Homeland Security Committee, Subcommittee on Cybersecurity, Infrastructure Protection, and Security Technologies, February 11, 2011.

[21] The DHS identified such iteration on the contents of site security plans as one factor delaying the start of the inspection process from December 2009 to February 2010. Oral testimony of Rand Beers, Under Secretary, National Protection and Programs Directorate, Department of Homeland Security, before the Senate Committee on Homeland Security and Governmental Affairs, March 3, 2010.

[22] Testimony of Rand Beers, Under Secretary, National Protection and Programs Directorate, Department of Homeland Security, before the House Committee on Energy and Commerce, Subcommittee on Environment and the Economy, March 31, 2011.

[23] Personal communication with Department of Homeland Security, September 15, 2011.

[24] Congressional policymakers have debated chemical facility security issues since at least the 106th Congress.

[25] House Committee on Homeland Security, Subcommittee on Transportation Security and Infrastructure Protection, *Chemical Security: The Implementation of the Chemical Facility Anti-Terrorism Standards and the Road Ahead*, 110th Congress, December 12, 2007.

[26] Department of Homeland Security, National Protection and Programs Directorate, Infrastructure Protection and Information Security, *Fiscal Year 2009 Congressional Justification*, p. IPIS-41.

[27] Congress required DHS in FY2006 and FY2007 to report on the resources needed to create and implement mandatory security requirements. See P.L. 109-295, Department of Homeland Security Appropriations Act, 2007, and H.Rept. 109-241, accompanying P.L. 109-90, Department of Homeland Security Appropriations Act, 2006.

[28] Oral testimony of Rand Beers, Under Secretary, National Protection and Programs Directorate, Department of Homeland Security, before the House Homeland Security Committee, Subcommittee on Cybersecurity, Infrastructure Protection, and Security Technologies, February 11, 2011.

[29] Monica Hatcher, "Why Chemical Plants Are Vulnerable to Terrorism," *Houston Chronicle*, April 5, 2010.

[30] For example, see Department of Homeland Security, *Chemical Facility Anti-Terrorism Standards Site Security Plans and Preliminary Inspections, NASTTPO Annual Meeting*, May 12, 2010; and W. Koch, Air Products, *Overview of DHS CFATS Pre Authorization Visit*, July 7, 2010.

[31] Personal communication with Department of Homeland Security, September 15, 2011.

[32] The DHS also states that it plans to inspect compliance at Tier 1 facilities annually (Department of Homeland Security, National Protection and Programs Directorate, Infrastructure Protection and Information Security, *Fiscal Year 2012 Congressional Justification*, p. 26).

[33] Department of Homeland Security, *The Chemical Facility Anti-Terrorism Standards—Update for the Chemical Sector Security Summit*, June 29, 2009.

[34] CRS calculation assuming two inspectors per inspection and one inspection per week.

[35] It should be noted that all facilities regulated under CFATS are by definition high-risk chemical facilities and that a lower or higher risk tier is relative to other high-risk chemical facilities.

[36] 72 *Federal Register* 17688–17745 (April 9, 2007) at 17739.

[37] Section 534, P.L. 110-161, Consolidated Appropriations Act, 2008.

[38] Several states, including New Jersey, Maryland, and New York, have implemented laws addressing security at chemical facilities.

[39] 72 *Federal Register* 17688–17745 (April 9, 2007) at 17727.

[40] See, for example, National Association of Chemical Distributors, "NACD Key Issue: Chemical Facility Security," *Key Issues 2009 Washington Fly-In 111th Congress*.

[41] For example, in the 111th Congress, Representative Rothman asked Secretary of Homeland Security Napolitano,

> And in particular, there was language enacted in 2008 which said that the states could have their own regulations with regard to securing chemical plant facilities unless there was a conflict with the federal requirements. Might it be time to revisit that language to allow each state to have its own chemical plant security regulations, even stricter than a national minimum standard, even if they conflict?

("House Appropriations Subcommittee on Homeland Security Holds Hearing on the Department of Homeland Security," *CQ Congressional Transcripts*, May 12, 2009.)

[42] OMB Watch and Public Citizen, "Chemical Facility Anti-Terrorism Standards, Department of Homeland Security, DHS-2006-0073," *Letter*, February 7, 2007.

[43] Testimony of Joseph Crawford, Chief of Police, City Saint Albans, West Virginia, before the House Committee on Energy and Commerce, Subcommittee on Oversight and Investigations, April 21, 2009; and testimony of Kent Carper, President, Kanawha County Commission, Kanawha County, West Virginia, before the House Committee on Energy and Commerce, Subcommittee on Oversight and Investigations, April 21, 2009.

[44] See, for example, testimony of Glenn Erwin, United Steelworkers International Union, before the Senate Committee on Homeland Security and Governmental Affairs, July 13, 2005.

[45] "House Energy and Commerce Subcommittee on Oversight and Investigations Holds Hearing on the Bayer CropScience Facility Explosion," *CQ Congressional Transcripts*, April 21, 2009.

[46] For example, see "House Energy and Commerce Subcommittee on Oversight and Investigations Holds Hearing on the Bayer CropScience Facility Explosion," *CQ Congressional Transcripts*, April 21, 2009.

[47] The DHS regulates for security purposes chemical facilities located in ports under the Maritime Transportation Security Act of 2002 (P.L. 107-295). The chemical facility security statute exempts chemical facilities regulated under MTSA.

[48] Testimony of William B. Buckner, President and Chief Executive Officer of Bayer CropScience, before the House Committee on Energy and Commerce, Subcommittee on Oversight and Investigations, April 21, 2009.

[49] 72 *Federal Register* 17688–17745 (April 9, 2007) at 17718.

[50] 72 *Federal Register* 65396–65435 (November 20, 2007) at 65406.

[51] 72 *Federal Register* 65396–65435 (November 20, 2007) at 65412.

[52] Congress required certain drinking water facilities to perform vulnerability assessments and develop emergency response plans through section 401 of P.L. 107-188, the Public Health Security and Bioterrorism Preparedness and Response Act of 2002. For more information on drinking water security activities, see CRS Report RL31294, *Safeguarding the Nation's Drinking Water: EPA and Congressional Actions*, by Mary Tiemann.

[53] See, for example, Paul Orum and Reece Rushing, Center for American Progress, *Chemical Security 101: What You Don't Have Can't Leak, or Be Blown Up by Terrorists*, November 2008; and testimony of Philip J. Crowley, Senior Fellow and Director of Homeland

Security, Center for American Progress, before the House Committee on Energy and Commerce, Subcommittee on Environment and Hazardous Materials, June 12, 2008.

[54] See U.S. Environmental Protection Agency, *Factoids: Drinking Water and Ground Water Statistics for 2008*, EPA 816-K-08-004, November 2008; and U.S. Environmental Protection Agency, *Clean Watersheds Needs Survey 2004: Report to Congress*, January 2008.

[55] American Water Works Association, "Chemical Facility Security," *Fact Sheet*, 2009, online at http://www.awwa.org/files/GovtPublicAffairs/PDF/2009Security.pdf. For more information on security issues in the water infrastructure sector, see CRS Report RL32189, *Terrorism and Security Issues Facing the Water Infrastructure Sector*, by Claudia Copeland.

[56] See U.S. Environmental Protection Agency, *Factoids: Drinking Water and Ground Water Statistics for 2008*, EPA 816-K-08-004, November 2008; and U.S. Environmental Protection Agency, *Clean Watersheds Needs Survey 2004: Report to Congress*, January 2008. For comparison, more than 38,000 chemical facilities filed a Top-Screen under CFATS.

[57] For example, the number of individuals served by the drinking water facility might be used as a regulatory criterion. Section 401 of the Public Health Security and Bioterrorism Preparedness and Response Act of 2002 (P.L. 107-188) mandated drinking water facilities serving more than 3,300 individuals develop an emergency response plan and perform a vulnerability assessment. Approximately 8,400 community water systems met this requirement at that time. For more information on drinking water security activities, see CRS Report RL31294, *Safeguarding the Nation's Drinking Water: EPA and Congressional Actions*, by Mary Tiemann.

[58] Oral testimony of Rand Beers, Under Secretary, National Protection and Programs Directorate, Department of Homeland Security, before the House Homeland Security Committee, Subcommittee on Cybersecurity, Infrastructure Protection, and Security Technologies, February 11, 2011.

[59] For example, the replacement of hydrogen fluoride with sulfuric acid for refinery processing would replace a more toxic chemical with a less toxic one. In this case, experts estimate that equivalent processing capacity would require twenty-five times more sulfuric acid. Thus, more chemical storage facilities and transportation would be required, potentially posing different dangers than atmospheric release to the surrounding community. Determining which chemical process had less overall risk might require considering factors both internal and external to the chemical facility and the surrounding community. See testimony of M. Sam Mannan, Director, Mary Kay O'Connor Process Safety Center, Texas A&M University, before the House Committee on Homeland Security, December 12, 2007.

[60] Testimony of M. Sam Mannan, Director, Mary Kay O'Connor Process Safety Center, Texas A&M University, before the House Committee on Homeland Security, December 12, 2007.

[61] Committee on Assessing Vulnerabilities Related to the Nation's Chemical Infrastructure, National Research Council, *Terrorism and the Chemical Infrastructure: Protecting People and Reducing Vulnerabilities*, 2006.

[62] For further discussion on this issue, see Center for Chemical Process Safety, American Institute of Chemical Engineers, *Final Report: Definition for Inherently Safer Technology in Production, Transportation, Storage, and Use,* July 2010.

[63] See, for example, Paul Orum and Reece Rushing, Center for American Progress, *Preventing Toxic Terrorism: How Some Chemical Facilities are Removing Danger to American Communities*, April 2006; and Paul Orum and Reece Rushing, Center for American Progress, *Chemical Security 101: What You Don't Have Can't Leak, or Be Blown Up by Terrorists*, November 2008.

[64] See, for example, testimony of Timothy J. Scott, Dow Chemical Company, before the House Committee on Homeland Security, Subcommittee on Cybersecurity, Infrastructure Protection, and Security Technologies, February 11, 2011; and testimony of Marty Durbin, Managing Director, Federal Affairs, American Chemistry Council, before the House

Committee on Energy and Commerce, Subcommittee on Environment and Hazardous Materials, June 12, 2008.

[65] AcuTech Consulting Group, *A Survey of CFATS Progress in Securing the Chemical Sector*, September 6, 2011, p. 41.

[66] For example, EPA experts have pointed to the change by drinking water treatment facilities between two approved disinfectants—chlorine and chloramine—as correlated with an unexpected increase in levels of lead in drinking water due to increased corrosion. Government Accountability Office, *Lead in D.C. Drinking Water*, GAO-05-344, March 2005.

[67] See, for example, testimony of M. Sam Mannan, Director, Mary Kay O'Connor Process Safety Center, Texas A&M University, before the House Committee on Homeland Security, Subcommittee on Cybersecurity, Infrastructure Protection, and Security Technologies, February 11, 2011; testimony of Dennis C. Hendershot, Staff Consultant, Center for Chemical Process Safety, American Institute of Chemical Engineers, before the Senate Committee on Environment and Public Works, June 21, 2006, S.Hrg. 109-1044; and testimony of Matthew Barmasse, Synthetic Organic Chemical Manufacturers Association, before the Senate Committee on Homeland Security and Governmental Affairs, July 13, 2005.

[68] Testimony of Stephen Poorman, International EHS Manager, FUJIFILM Imaging Colorants Ltd., on behalf of the Society of Chemical Manufacturers and Affiliates before the Senate Committee on Homeland Security and Governmental Affairs, March 3, 2010.

[69] 72 *Federal Register* 17688–17745 (April 9, 2007) at 17772.

[70] Department of Homeland Security, *Chemical Facility Anti-Terrorism Standards Interim Final Rule Regulatory Assessment,* DHS-2006-0073, April 1, 2007.

[71] Office of the General Counsel, General Accounting Office, *Principles of Federal Appropriations Law, Third Edition*, GAO-04-261SP, January, 2004, pp. 2-70–2-71.

[72] Office of Management and Budget, The White House, *Budget of the United States Government, Fiscal Year 2012, Appendix*, p. 553.

[73] Oral testimony of Rand Beers, Under Secretary, National Protection and Programs Directorate, Department of Homeland Security, before the House Homeland Security Committee, Subcommittee on Cybersecurity, Infrastructure Protection, and Security Technologies, February 11, 2011.

[74] Randy Dearth and Cal Dooley, "Commentary: Taking Chemical Plant Security In Pittsburgh Seriously," *Pittsburgh Post-Gazette*, May 27, 2009.

[75] 72 *Federal Register* 17688–17745 (April 9, 2007) at 17712.

[76] 73 *Federal Register* 1640 (January 9, 2008).

[77] Testimony of Benjamin H. Grumbles, Assistant Administrator for Water, U.S. Environmental Protection Agency before the House Committee on Energy and Commerce, Subcommittee on Environment and Hazardous Materials, June 12, 2008. See also testimony of Rand Beers, Under Secretary, National Protection and Programs Directorate, Department of Homeland Security, before the Senate Committee on Homeland Security and Governmental Affairs, March 3, 2010.

[78] Testimony of Peter S. Silva, Assistant Administrator for Water, Environmental Protection Agency, before the Senate Committee on Homeland Security and Governmental Affairs, March 3, 2010.

[79] Testimony of Peter S. Silva, Assistant Administrator for Water, Environmental Protection Agency, before the Senate Committee on Homeland Security and Governmental Affairs, March 3, 2010.

[80] Testimony of Rand Beers, Under Secretary, National Protection and Programs Directorate, Department of Homeland Security, before the Senate Committee on Homeland Security and Governmental Affairs, March 3, 2010. The DHS and the Nuclear Regulatory Commission have developed a memorandum of agreement regarding security at chemical facilities regulated by the Nuclear Regulatory Commission (*Memorandum of*

Understanding between the U.S. Department of Homeland Security and the U.S. Nuclear Regulatory Commission, March 31, 2011).

[81] Those chemical facilities exempt from CFATS because they are regulated under MTSA are overseen by the Coast Guard, which is part of DHS. The DHS testified that 365 facilities are fully exempt from CFATS regulation due to compliance with MTSA, while 135 are partially exempt ("House Homeland Security Committee Holds Hearing on the Chemical Facility Antiterrorism Act of 2009," *CQ Congressional Transcripts*, June 16, 2009).

[82] Section 1433 of the Safe Drinking Water Act as amended by section 401 of P.L. 107-188, the Public Health Security and Bioterrorism Preparedness and Response Act of 2002.

[83] Testimony of Brad Coffey, Association of Metropolitan Water Agencies, before the House Committee on Energy and Commerce, Subcommittee on Environment and Hazardous Materials, June 12, 2008.

[84] See, for example, American Water Works Association, "AWWA Members Urged to Contact Congress on Chemical Security Bill," and Association of Metropolitan Water Agencies, "Drinking Water Security and Treatment Mandates," *Policy Resolution*, October 2008.

[85] Some agencies oversee both safety and security issues. For example, the U.S. Coast Guard has both safety and security responsibilities for ports.

[86] Testimony of Brad Coffey, Association of Metropolitan Water Agencies, before the House Committee on Energy and Commerce, Subcommittee on Environment and Hazardous Materials, June 12, 2008.

[87] Testimony of Rand Beers, Under Secretary, National Protection and Programs Directorate, Department of Homeland Security, before the Senate Committee on Homeland Security and Governmental Affairs, March 3, 2010.

[88] Testimony of Peter S. Silva, Assistant Administrator for Water, Environmental Protection Agency, before the Senate Committee on Homeland Security and Governmental Affairs, March 3, 2010.

[89] Testimony of Rand Beers, Under Secretary, National Protection and Programs Directorate, Department of Homeland Security, before the House Committee on Energy and Commerce, Subcommittee on Environment and the Economy, March 31, 2011.

[90] Testimony of Rand Beers, Under Secretary, National Protection and Programs Directorate, Department of Homeland Security, before the Senate Committee on Homeland Security and Governmental Affairs, March 3, 2010.

[91] See, for example, testimony of Dennis C. Hendershot, Staff Consultant, Center for Chemical Process Safety, American Institute of Chemical Engineers, before the Senate Committee on Environment and Public Works, June 21, 2006, S.Hrg. 109-1044.

[92] The DHS Science and Technology (S&T) Directorate is engaged in a Chemical Infrastructure Risk Assessment Project that, among other goals, will assess the potential for safer alternative processes that may reduce risk to a select subset of high volume toxic chemicals (Department of Homeland Security, *FY2010 Budget Justification*, pp. S&T R&D - 27–28). The Chemical Security Analysis Center of the DHS S&T Directorate contracted with the Center for Chemical Process Safety of the American Institute of Chemical Engineers to develop a technically based definition for inherently safer technology. See Center for Chemical Process Safety, American Institute of Chemical Engineers, *Final Report: Definition for Inherently Safer Technology in Production, Transportation, Storage, and Use*, July 2010.

[93] Section 401 of the Public Health Security and Bioterrorism Preparedness and Response Act of 2002 (P.L. 107-188) mandated drinking water facilities serving more than 3,300 individuals develop an emergency response plan and perform a vulnerability assessment. Funds were authorized to help offset the costs to these facilities.

[94] Following investigation into the explosion at the Bayer CropScience facility in Institute, WV, Members of Congress requested that the Chemical Safety Board provide recommendations on the adoption of alternative chemical processes at the chemical facility. Rep. Henry A. Waxman, Sen. John D. Rockefeller IV, Rep. Bart Stupak, and Rep. Edward J. Markey,

Letter to John Bresland, May 4, 2009, online at http://energycommerce.house.gov/Press_111/20090504/bayer.pdf.

[95] Testimony of Rand Beers, Under Secretary, National Protection and Programs Directorate, Department of Homeland Security, before the Senate Committee on Homeland Security and Governmental Affairs, March 3, 2010.

[96] A similar approach was taken with regard to making available chemical facility information submitted to the EPA under the auspices of the Risk Management Program. In this case, Congress directed the President to assess the potential risk of placing this information on the Internet. See Section 3 of Chemical Safety Information, Site Security and Fuels Regulatory Relief Act (P.L. 106-40). See Department of Justice, *Assessment of the Increased Risk of Terrorist or Other Criminal Activity Associated with Posting Off-Site Consequence Analysis Information on the Internet*, April 18, 2000.

[97] While DHS has established a "CFATS Tip-Line" where individuals may report security concerns, individuals using the tip-line accrue no special protections.

[98] P.L. 110-161, the Consolidated Appropriations Act, 2008, Section 534.

[99] Office of Management and Budget, The White House, *Budget of the United States Government, Fiscal Year 2011, Appendix*, p. 574; Office of Management and Budget, The White House, *Budget of the United States Government, Fiscal Year 2012, Appendix*, p. 553.

In: Chemical Facility Security
Editors: M. J. Flores et al.

ISBN 978-1-61942-925-3

Chapter 2

CHEMICAL FACILITY SECURITY HEARING—LUNGREN STATEMENT*

Dan Lungren

I want to welcome our witnesses today to our first Subcommittee hearing under the new Republican majority. I also want to thank the Ranking Member Ms Clarke, for the bi-partisan working relationship we enjoyed in the 111th Congress and I will continue that bi-partisan spirit in the 112th Congress. I'm very happy to welcome my new Republican colleagues (Mr. McCaul is now a veteran) to our subcommittee, Mr. Walberg, Mr. Meehan, Mr. Long and Mr. Marino. Our subcommittee will be examining many critical issues in this Congress from physical and cyber threats to our critical infrastructure, to radinuc/bio threats to our cities. We will be aggressive in our oversight of the department and its many security programs, especially those that provide "no substantial security improvement" for the amount of taxpayer dollars spent.

One of those critical issues in our jurisdiction is chemical security. Chemicals underpin our way of life and our 21st-century economy by employing over 800,000 workers who produce 19% of the world's chemical products. Because of this critical economic role of the chemical industry and the danger of these chemicals for facility workers and surrounding populations

---

* These remarks were given as a statement before the Subcommittee on Cybersecurity, infrastructure Protection and Security Technology Hearing on February 22, 2011. Dan Lungren is Chairman of the subcommittee.

due to the terrorist threat, securing chemical facilities is a top priority for our committee.

My interest in this issue dates back to the last time I was chairman of this subcommittee and I introduced the Chemical Security Anti-Terrorism Act of 2006. At that time, the chemical industry was operating under a voluntary security regime which left many security gaps because of non-participating facilities. In order to address those security gaps, I introduced my bill which would have established a risk-based performance approach targeting high risk chemical facilities. While the bill was never considered by the House, it provided the model used by the administration and house appropriators to craft a compromise national risk- based security plan for all high risk chemical facilities (Sec. 550 of the Homeland Security Appropriations Act of 2007) in the closing weeks of the 109th Congress.

The authority to regulate chemical facility security is historic and critically important. Dangerous chemicals, listed in Appendix A as chemicals of interest (COI) when stored or processed above threshold quantities, pose serious threats to facility workers and neighboring populations. They are also attractive terrorist targets. We need to bolster security at chemical facilities and the best way to do that is to allow the CFATS (Chemical Facility Anti Terrorism.) Program to be fully implemented before any significant program changes are enacted.

The best way to strengthen the foundation of security that CFATS is building at our nation's chemical facilities is to provide a "long-term extension of the CFATS authority". This will provide our chemical facility partners, who are spending millions of dollars implementing new security measures, with the assurance that the rules and requirements won't change from year to year as the CFATS program is implemented. This will also provide the department with the certainty that Congress believes chemical security will be a priority for years to come as they continue to implement, evolve, and invest in the CFATS program.

Another controversial issue that has emerged during chemical security debates centers on the understanding of "inherently safer technology" or 1ST. I don't support mandating single solution security approaches. 1ST isn't something you can buy off the shelf or simply plug in. It is a concept, a not very well understood concept at that. It is a very complex process or series of procedures that should not be mandated, according to the testimony of our three non-governmental witnesses today. There is no single definition of what 1ST "is" — because it differs so greatly from chemical to chemical, and from facility to facility.

I support risk-based security solutions that reduce identified vulnerabilities and would oppose mandating specific security measures. Requiring a specific type of security measure goes against the very principles of a risk-informed, performance- based approach. CFATS is building a strong chemical security foundation by enabling multiple risk-based solutions and flexibility for facilities to select the security approaches that best fit their unique security needs while still meeting the risk-based performance standards established by DHS.

The original bi-partisan Congressional goal of these regulations was to strike the right balance between improving security at our high risk chemical facilities while preserving the economic vitality of this critical sector. While CFATS still has a long way to go, I believe that the CFATS authority enabling the regulatory structure in use today, is providing our Nation the best opportunity to meet that Congressional goal.

I want to thank our witnesses for appearing today and I look forward to hearing your testimony.

In: Chemical Facility Security
Editors: M. J. Flores et al.
ISBN 978-1-61942-925-3

Chapter 3

CHEMICAL FACILITY SECURITY HEARING—BEERS TESTIMONY*

Rand Beers

Thank you, Chairman Lungren, Ranking Member Clarke, and distinguished members of the subcommittee. It is a pleasure to appear before you today to discuss the Department of Homeland Security's (DHS) efforts to secure high-risk chemical facilities. As you are aware, the department's current authority under Section 550 of the Fiscal Year 2007 Department of Homeland Security Appropriations Act, as amended, was set to expire in October 2010, but has been temporarily extended under the current Continuing Resolution. DHS is eager to work with this committee, Congress, and all levels of government and the private sector to achieve passage of legislation that permanently authorizes and appropriately matures the Chemical Facility Anti- Terrorism Standards (CFATS) program. In the interest of facilitating that collaboration, my testimony focuses on the current program and the key principles that DHS would like to see guide the program's maturation.

* These remarks were given as a testimony on February 11, 2011. Rand Beers is the Under Secretary National Protection and Programs Directorate at the Department of Homeland Security.

Chemical Security Regulations

Section 550 of the Fiscal Year (FY) 2007 Department of Homeland Security Appropriations Act directed the department to develop and implement a regulatory framework to address the high level of security risk posed by certain chemical facilities. Specifically, Section 5 50(a) of the Act authorized the department to adopt rules requiring high-risk chemical facilities to complete Security Vulnerability Assessments (SVAs), develop Site Security Plans (SSPs), and implement protective measures necessary to meet risk-based performance standards established by the department. Consequently, the department published an Interim Final Rule, known as CFATS, on April 9, 2007. Section 550, however, expressly exempts from those rules certain facilities that are regulated under other federal statutes, including those regulated by the United States Coast Guard pursuant to the Maritime Transportation Security Act (MTSA), drinking water and wastewater treatment facilities as defined by Section 1401 of the Safe Water Drinking Act and Section 212 of the Federal Water Pollution Control Act, and facilities owned or operated by the departments of Defense and Energy, as well as certain facilities subject to regulation by the Nuclear Regulatory Commission (NRC).

The following core principles guided the development of the CFATS regulatory structure:

1) *Securing high-risk chemical facilities is a comprehensive undertaking that involves a national effort, including all levels of government and the private sector.* Integrated and effective participation by all stakeholders—federal, state, local, tribal and territorial government partners as well as the private sector—is essential to securing our critical infrastructure, including high-risk chemical facilities. Implementing this program means tackling a sophisticated and complex set of issues related to identifying and mitigating vulnerabilities and setting security goals. This requires a broad spectrum of input, as the regulated facilities bridge multiple industries and critical infrastructure sectors. By working closely with experts, members of industry, academia, and federal government partners, we leveraged vital knowledge and insight to develop the regulation.
2) *Risk-based tiering to guide resource allocations.* Not all facilities present the same level of risk. The greatest level of scrutiny should be

focused on those facilities that present the highest risk—those that, if attacked, would endanger the greatest number of lives.

3) *Reasonable, clear, and calibrated performance standards will lead to enhanced security.* The current CFATS rule includes enforceable risk-based performance standards. High- risk facilities have the flexibility to develop appropriate site-specific security measures that will effectively address risk. The department will analyze each final tiered facility's SSP to see if it meets CFATS performance standards. If necessary, DHS will work with the facility to revise and resubmit an acceptable plan.
4) *Recognition of the progress many companies have already made in improving facility security leverages those advancements.* Many companies have made significant capital investments in security since 9/11. Building on that progress in implementing the CFATS program will raise the overall security baseline at high-risk chemical facilities.

On Nov. 20, 2007, the department published Appendix A to CFATS, which lists 322 chemicals of interest—including common industrial chemicals such as chlorine, propane, and anhydrous ammonia—as well as specialty chemicals, such as arsine and phosphorus trichloride. The department included chemicals based on the consequences associated with one or more of the following three security issues:

1) Release – Toxic, flammable, or explosive chemicals that have the potential to create significant adverse consequences for human life or health if intentionally released or detonated;
2) Theft/Diversion – Chemicals that have the potential, if stolen or diverted, to be used or converted into weapons that could cause significant adverse consequences for human life or health; and
3) Sabotage/Contamination – Chemicals that, if mixed with other readily available materials, have the potential to create significant adverse consequences for human life or health.

The department also established a Screening Threshold Quantity for each chemical of interest based on its potential to create significant adverse consequences to human life or health in one or more of these ways.

Implementation and execution of the CFATS regulation requires the department to identify which facilities it considers high-risk. The department developed the Chemical Security Assessment Tool (CSAT) to identify potentially high-risk facilities and to provide methodologies that facilities can use to conduct SVAs and to develop SSPs. CSAT is a suite of online applications designed to facilitate compliance with the program; it includes user registration, the initial consequence-based screening tool (Top-Screen), an SVA tool, and an SSP template. Through the Top-Screen process, the department initially identifies and sorts facilities based on their associated risks.

If a facility is initially identified during the Top-Screen process as potentially having a level of risk subject to regulation under CFATS, the department assigns the facility to one of four preliminary risk-based tiers, with Tier 1 representing the highest level of potential risk. Those facilities must then complete SVAs and submit them to the department, although facilities preliminarily designated as Tier 4 facilities also have the option of submitting an Alternative Security Program (ASP). Results from the SVA inform the department's final determinations as to whether a facility is in fact high-risk and, if so, of the facility's final tier assignment. Each SVA is carefully reviewed for its description of how chemicals of interest are actually held at the site, how those chemicals are managed, and for physical, cyber, and chemical security risks.

After completing its review of a facility's SVA, the department makes a final determination as to whether the facility is considered high-risk and assigns the facility a final risk-based tier. Final high-risk facilities are then required to develop an SSP or, if they so choose, an ASP that addresses its identified vulnerabilities and security issues. Only facilities that receive a final high-risk determination letter under CFATS will be required to complete and submit an SSP or, if the facility so chooses, an ASP. DHS' final determinations of which facilities are high-risk are based on each facility's individual consequentiality and vulnerability as determined by its Top- Screen, SVA, and any other available information. The higher the facility's risk-based tier, the more robust the security measures and the more frequent and rigorous the inspections will be. The purpose of inspections is to validate the adequacy of a facility's SSP and to verify that measures identified in the plan are being implemented.

IMPLEMENTATION STATUS

To date, the department has reviewed more than 39,000 Top-Screen consequence assessment questionnaires submitted by potentially high-risk chemical facilities. Since June 2008, we have notified more than 7,000 preliminarily tiered facilities that they have been initially designated as high-risk and are thus required to submit SVAs; we have nearly completed our review of the almost 6,200 SVAs that have been submitted. In May 2009, we began notifying facilities of their final high-risk determinations, risk-based tiering assignments, and the requirement to complete and submit an SSP or ASP.

In May 2009, the department issued 141 final tier determination letters to the highest risk (Tier 1) facilities, confirming their high-risk status and initiating the 120-day time frame for submitting an SSP. After issuing this initial set of final tier determinations, the department periodically issued notifications to additional facilities of their final high-risk status. To date, more than 4,000 additional facilities have received final high-risk determinations and tier assignments, and several hundred that were preliminarily tiered by DHS were informed that they are no longer considered high-risk.

CFATS currently covers 4,755 high-risk facilities nationwide across all 50 states, of which 4,094 facilities have received final high-risk determinations and due dates for submission of an SSP or ASP. More than 4,000 facilities have submitted SSPs (or ASPs) to date, and the department is in the process of reviewing these submissions. The department continues to issue final tier notifications to facilities across all four risk tiers as additional final tier determinations are made by the department.

In February 2010, the department began conducting inspections of final-tiered facilities, starting with the Tier 1-designated facilities, and has completed more than 175 pre-authorization inspections to date. The department intends to use these initial inspections to help gain a comprehensive understanding of the processes, risks, vulnerabilities, response capabilities, security measures and practices, and any other factors that may be in place at a regulated facility that affect security risk in order to help facilities submit SSPs that can be approved under CFATS. After DHS issues a letter of authorization for a facility's SSP, DHS will conduct a comprehensive and detailed compliance inspection before making a final determination as to whether the facility has appropriately enacted their SSP. Facilities that have successfully implemented their approved SSPs and have passed an inspection will be considered in compliance with the required performance standards.

A critical element of the department's efforts to secure the nation's high-risk chemical facilities, the SSP enables final high-risk facilities to document their individual security strategies for meeting the Risk-Based Performance

Standards (RBPS) established under CFATS. Each high- risk facility's security strategy will be unique, as it depends on the facility's risk level, security issues, characteristics, and other factors. Therefore, the SSP tool collects information on each of the 18 RBPS for each facility. The RBPS cover the fundamentals of security, such as restricting the area perimeter, securing site assets, screening and controlling access, cybersecurity, training and response. The SSP tool is designed to take into account the complicated nature of chemical facility security and allows facilities to describe both facility-wide and asset-specific security measures. The department understands that the private sector generally, and CFATS-affected industries in particular, are dynamic. The SSP tool allows facilities to involve their subject- matter experts from across the facility, company and corporation, as appropriate, in completing the SSP and submitting a combination of existing and planned security measures to satisfy the RBPS. The department expects that most approved SSPs will consist of a combination of existing and planned security measures. Through a review of the SSP, in conjunction with an on-site inspection, DHS will determine whether a facility has met the requisite level of performance given its risk profile and thus whether its SSP should be approved.

Along with the initial group of final Tier 1 notifications and the activation of the SSP tool in May 2009, DHS issued the *Risk-Based Performance Standards Guidance* document. The department developed this guidance to assist high-risk chemical facilities subject to CFATS in determining appropriate protective measures and practices to satisfy the RBPS. It is designed to help facilities comply with CFATS by providing detailed descriptions of the 18 RBPS as well as examples of various security measures and practices that could enable facilities to achieve the appropriate level of performance for the RBPS at each tier level. The *Guidance* also reflects public and private sector dialogue on the RBPS and industrial security, including public comments on the draft guidance document. High-risk facilities are free to make use of whichever security programs or processes they choose—whether or not in the *Guidance*— provided that they achieve the requisite level of performance under the CFATS RBPS. The *Guidance* will, however, help high-risk facilities gain a sense of what types and combination of security measures may satisfy the RBPS. The department has also offered regular SSP training webinars to assist high-risk facilities with completing their SSPs.

For additional context, I would like to provide you with an example of how some facilities may be approaching the development and submission of their SSPs: in the case of a Tier 1 facility with a release hazard security issue, the facility is required to restrict the area perimeter appropriately, which may include preventing breach by a wheeled vehicle. To meet this standard, the facility is able to consider numerous security measures, such as cable anchored in concrete block along with movable bollards at all active gates or perimeter landscaping (e.g., large boulders, steep berms, streams, or other obstacles) that would thwart vehicle entry. The department will approve the security measure as long as it is determined by the department to be sufficient to address the applicable performance standard. Under Section 550, the department cannot mandate a specific security measure to approve the SSP.

In June 2010, the department issued its first Administrative Orders under CFATS to 18 chemical facilities for failure to submit an SSP. Throughout the remainder of the year, the department issued an additional 47 Administrative Orders to chemical facilities that had failed to submit an SSP in a timely manner. Administrative Orders are the first step toward enforcement under CFATS. An Administrative Order does not impose a penalty or fine, but directs the facility to take specific action to comply with CFATS—in this case, to complete the SSP within 10 days of receipt. If the facility does not comply with the Administrative Order, however, the department may issue an Order Assessing Civil Penalty of up to $25,000 each day the violation continues, or an Order to Cease Operations. All 65 facilities that received an Administrative Order ultimately completed their SSPs following receipt of the Administrative Order, or providing amplifying information to the department, that satisfactorily explained why they had failed to meet the deadline for submitting their SSPs, and thus, no further enforcement action was necessary. As CFATS implementation progresses, the department expects to continue to exercise its enforcement authority to ensure CFATS compliance.

OUTREACH EFFORTS

Since the release of CFATS in April 2007, the department has taken significant steps to publicize the rule and ensure that the regulated community and our security partners are aware of its requirements. As part of this outreach program, the department has regularly updated impacted sectors through their Sector Coordinating Councils and the Government Coordinating Councils of industries most impacted by CFATS, including the Chemical, Oil and Natural Gas, and Food and Agriculture Sectors. We have also solicited feedback from our public and private sector partners and, where appropriate, have reflected that feedback in our implementation activities. As the program continues to mature, the department participates in an average of 250 CFAT S-specific outreach engagements annually, not including formal coordination activities with individual facilities such as pre-authorization inspections and Compliance Assistance Visits. We have presented at numerous security and chemical industry conferences; participated in a variety of other meetings of relevant security partners; established a Help Desk for CFATS questions that receives between 40 and 80 calls daily; put in place a CFATS tip-line for anonymous chemical security reporting; and developed and regularly updated a highly regarded Chemical Security website (www.DHS.gov/chemicalsecurity). This month, the department updated the CFATS website to include a more robust, searchable Knowledge Center, which further supports the regulated community. These efforts are having a positive impact: again, more than 39,000 Top-Screens have been submitted to the department via CSAT.

In addition, the department continues to focus on fostering solid working relationships with state and local officials as well as first responders in jurisdictions with high-risk facilities. To meet the risk-based performance standards under CFATS, facilities need to cultivate and maintain effective working relationships—including a clear understanding of roles and responsibilities— with local officials who aid in preventing, mitigating and responding to potential attacks. To facilitate these relationships, our inspectors have been actively working with facilities and officials in their areas of operation, and they have participated in more than 100 Local Emergency Planning Committee meetings to provide a better understanding of CFATS requirements. Last year, the department, in collaboration with the State, Local, Tribal, and Territorial Government Coordinating Council, issued a tri-fold brochure which summarizes CFATS programs and processes for local emergency responders.

In May 2010, the department launched a web-based information-sharing portal called "CFATSShare." This tool provides interested state Homeland Security Advisors, DHS Protective Security Advisors, and fusion centers access to detailed CFATS facility information as needed. In the future, DHS plans to make this tool available to other federal security partners, such as the Federal Bureau of Investigation. The department continues to improve the CFAT S-Share web portal based on feedback from users.

Additionally, the department continues to actively collaborate across components within DHS and with other federal agencies in the area of chemical security, including routine coordination between the department's National Protection and Programs Directorate (NPPD) and the United States

Coast Guard (USCG), the Transportation Security Administration, the Department of Justice's Federal Bureau of Investigation and Bureau of Alcohol, Tobacco, Firearms and Explosives, the NRC, and the Environmental Protection Agency (EPA). One primary example of this coordination includes the establishment of a joint NPPD/USCG CFATS-MTSA Working Group to evaluate and, where appropriate, implement methods to harmonize the CFATS and MTSA regulations. Similarly, the department has been working closely with the EPA to begin evaluating how the CFATS approach could be used for water and wastewater treatment facilities, should the water and wastewater treatment facility exemption be removed by Congress in future versions of chemical facility security or water facility security regulations.

The department also launched an Agricultural Facility Survey in July 2010. The goal of the survey is to provide the department with additional information on the potential risks related to agricultural Chemicals of Interest throughout the distribution chain—including manufacturers, distributors, retailers, commercial applicators, and end-users. The survey results will also help the department determine the most appropriate approach for addressing the existing extension of the CFATS Top Screen due date for agricultural production facilities. The department received completed surveys from nearly 1,200 CFATS facilities and is currently analyzing the results to determine the best approach to take regarding agricultural production facilities.

Internally, we are continuing to build the Infrastructure Security Compliance Division that is responsible for implementing CFATS. We have hired, or are in the process of on-boarding, more than 178 people, and we are continuing to hire throughout this fiscal year to meet our staffing goal of 268 positions. These numbers include our field inspector cadre, where we have hired 95 of 103 field inspector positions and 14 of 14 field leadership positions.

Legislation to Permanently Authorize CFATS

We have enjoyed the constructive dialogue with Congress, including Members of this Committee, as it contemplates new authorizing legislation. The department recognizes the significant work that this Committee and others, including the Senate Committee on Homeland Security and Governmental Affairs, the Senate Committee on Environment and Public Works and the House Committee on Energy and Commerce, have completed in reauthorizing the CFATS program to date and to address chemical facility security. We appreciate this effort and look forward to continuing the constructive engagement with Congress on these important matters.

The department supports a permanent authorization for the CFATS program. The department is committed to working with Congress and other security partners to pass stand-alone chemical security legislation that includes permanent authority beginning in FY 2011. The latest Continuing Resolution authorizes an extension of the statutory authority for CFATS, which otherwise would have sunset on Oct. 4, 2010.

It is important to highlight that the Administration has developed a set of guiding principles for the reauthorization of CFATS. These principles are the foundation for the department's position on permanent CFATS reauthorization:

- The Administration supports permanent authorization to regulate security of high-risk chemical facilities through risk-based performance standards.
- The department should be given reasonable deadlines by Congress to promulgate new rules to implement any new legislative requirements. CFATS, as currently being implemented, should remain in effect until or unless it is supplemented by new regulations.
- The Administration supports, where possible, using safer technology, to enhance the security of the nation's high-risk chemical facilities. Similarly, we recognize that risk management requires balancing threat, vulnerabilities, and consequences with the costs and benefits of mitigating risk. In this context, the Administration has established the following policy principles in regard to inherently safer technologies (IST) at high-risk chemical facilities:
    - The Administration supports consistency of IST approaches for facilities regardless of sector.
    - The Administration believes that all high-risk chemical facilities, Tiers 1-4, should assess IST methods and report the assessment in the facilities' SSPs.
    - Further, the appropriate regulatory entity should have the authority to require facilities posing the highest degree of risk (Tiers 1 and 2) to implement IST method(s) if such methods demonstrably enhance overall security, are determined to be feasible, and, in the case of water sector facilities, consider public health and environmental requirements.

- For Tier 3 and 4 facilities, the appropriate regulatory entity should review the IST assessment contained in the SSP. The entity should be authorized to provide recommendations on implementing IST, but it would not have the authority to require facilities to implement the IST methods.
- The Administration believes that flexibility and staggered implementation would be required in implementing this new IST policy.

- The Administration supports maintaining the department's current Chemical-terrorism Vulnerability Information regime for protecting sensitive information relating to chemical facility security. This regime is similar to, but distinct from, other Controlled Unclassified Information protection regimes.
- The department supports amending the current exemption for drinking water and wastewater facilities to specify that the Environmental Protection Agency (EPA) would have the lead on regulating for security, with DHS supporting EPA to ensure consistency across all sectors. This consistency could be achieved, for example, by the use of CFATS compliance tools and risk analysis with modifications as necessary to reflect the uniqueness of the water sector and statutory requirements. As DHS and EPA have stated before, we believe that there is a critical gap in the U.S. chemical facility security regulatory framework—namely, the exemption of drinking water and wastewater treatment facilities from CFATS. We need to work with Congress to close this gap to secure chemicals of interest at these facilities and to protect the communities that they serve; drinking water and wastewater treatment facilities that meet CFATS thresholds for chemicals of interest should be regulated. We do, however, recognize the unique public health and environmental requirements and responsibilities of such facilities. For example, we understand that a cease-operations order that might be appropriate for another facility under CFATS would have significant public health and environmental consequences when applied to a water facility.
- As you are aware, facilities regulated under MTSA authority are statutorily exempted from CFATS and thus are not required to submit Top-Screens to DHS. In order to help DHS develop a more comprehensive picture of security issues at the nation's chemical facilities, and to help DHS evaluate whether any regulatory gaps exist that may pose an unacceptable security risk, the department has begun

the process, with close cooperation between NPPD and USCG, for determining whether and how to require MTSA-covered facilities that possess CFATS chemicals of interest to complete and submit CFATS Top- Screens.

- With respect to the other current statutory exemptions, the department supports:
  - Maintaining the exemptions for both Defense and Energy Department facilities. Although the Department of Energy is exempt from the current statute, DOE policy does require chemical sabotage assessments utilizing the select agents lists and the adoption of protection measure where necessary; and
  - Amending the exemption for facilities regulated under the NRC to clarify the scope of the NRC exemption and to specify that DHS and the NRC shall work together to make a final determination on whether a facility or an area within a facility is subject to NRC regulation and is thus exempt from DHS regulation.

Given the complexity of chemical facility regulation, implementation logistics, and resource implications, any requirements considered in prospective legislation should also be taken into account to avoid having the department extensively revisit aspects of the program that are either currently in place or which will be implemented in the near future.

Conclusion

The department is collaborating extensively with the public, including members of the chemical sector and other interested groups, to work toward our collective goals under the CFATS regulatory framework. In many cases, industry has voluntarily made tremendous progress to ensure the security and resiliency of its facilities and systems. As we implement CFATS, we will continue to work with industry, our federal partners, states, and localities to get the job done.

The Administration recognizes that CFATS reauthorization requires further technical work. The department is ready to engage in technical discussions with Committee staff, affected stakeholders, and others to work out the remaining details. We must focus our efforts on implementing a risk-

and performance-based approach to regulation and, in parallel fashion, continue to pursue the voluntary programs that have already resulted in considerable success. We look forward to collaborating with the Committee, industry, and government partners to ensure that the chemical facility security regulatory effort achieves success in reducing risk in the chemical sector.

Thank you for holding this important hearing. I would be happy to respond to any questions you may have.

In: Chemical Facility Security
Editors: M. J. Flores et al.
ISBN 978-1-61942-925-3

Chapter 4

CHEMICAL FACILITY SECURITY HEARING—SCOTT TESTIMONY*

Timothy J. Scott

Chairman Lungren, ranking member Clarke and members of the committee, I'm Timothy Scott, Chief Security Officer for the Dow Chemical Company. Dow is a member of the American Chemistry Council, and I'm speaking today on behalf of Dow and our industry association.

I would like to make three key points in my statement today that I think will demonstrate that we have a strong foundation in place to build from when it comes to securing the nation's chemical facilities:

First – The chemical industry is critical to our national economy as well as the quality of life of our people. We recognize the historical and current risks to our industry and have taken aggressive action to improve our security posture – voluntarily well before the creation of DHS and the Chemical Facility Anti-Terrorism Standards (CFATS) and continuing now with the ongoing implementation of CFATS.

Second - The implementation of CFATS to date is demonstrably achieving its objectives to reduce the number of high risk sites and lower the risk profile of remaining high risk sites.

---

* These remarks were given as a testimony on February 11, 2011. Timothy J. Scott is Chief Security Officer & Corporate Director, Emergency Services & Security at the Dow Chemical Company.

And third – and most important – we need to continue on this successful path, maintain the momentum we've achieved and complete the task at hand without further delays or calls for change that would result in slowing the progress being made on chemical facility security by both DHS and the chemical industry..

Last month, President Obama issued an Executive Order which lays out a set of regulatory principles. These principles call for a system that promotes "economic growth, innovation, competiveness and job creation." They also call for regulations that promote predictability, and use the "best, most innovative and least burdensome tools for achieving regulatory ends."

We couldn't agree more that we need strong, sound, efficient regulations that will not hinder the ability of American companies to compete and create jobs.

This is particularly important for an industry like ours. Chemistry is the source of many of the new technologies that will help create jobs in the future, drive economic growth and achieve the goals articulated by the president during his State of the Union including clean energy; improved infrastructure; efficient transportation options; medical advancements that bring down the cost of health care; and even a strong defense.

And the Business of Chemistry employs nearly 800,000 people directly in high-paying, high-skill jobs. These are the kind of jobs that not only put food on the table, but boost consumer spending, send kids to college, allow families to own homes, and save for retirement.

Because of our critical role in the economy and our responsibility to our employees, communities and shareholders - safety and security continues to be a top priority for Dow and the members of ACC. And, we have actively demonstrated the commitment to security. Throughout our history, the chemical industry has implemented innovative processes and technologies to enhance safety. Staring from the design and construction phase of new facilities – by far the most efficient and logical time to achieve the maximum value and process safety improvements – to measures that are implemented throughout the distribution chain, our focus on security is clear and evident.

In 2001, ACC members voluntarily adopted an aggressive security program – already in draft form before 9/11 -- that became the Responsible Care® Security Code (RCSC). The Security Code is a comprehensive security program that addresses both physical and cyber security vulnerabilities, and requires ACC members to perform an extensive assessment of its security risks. Implementation of the code and regular independent review is mandatory for membership in the ACC. In fact, the strength of the code has been recognized by DHS and is designated as a qualified anti-terrorism technology under the Safety Act.

Recognizing the important role of government to address the risk of and thwart terrorism, Dow, along with ACC, has been proactive in our support for a national regulatory program to address security across the chemical sector. We strongly supported the bipartisan legislation that gave DHS the authority to regulate chemical security and create the Chemical Facility Anti-Terrorism Standards (CFATS).

CFATS is by far the most robust, comprehensive and demanding chemical security regulatory program to date. It takes a well-designed risk-based approach, sets a high bar through performance-based standards, and holds high risk facilities accountable for meeting those standards.

The accountability within the program drives facilities to consider all potential risk- reduction options, including potential process safety improvements, when developing a site security plan. Just as important, it leaves the decision of how to meet the standards to the site's discretion and subject to DHS approval of the site security plan. The result is a security plan approved by DHS that is uniquely and appropriately designed by the site to address the specific risk issues of each individual facility and meet the performance standards of DHS.

While still in the initial implementation stages, these tough regulations are yielding measurable results to manage or in some cases eliminate security risks. DHS has screened chemical facilities across the country and identified all of the facilities it deems are a "high-risk" for a potential terrorist attack. These facilities have been notified of their regulatory obligations and many have already taken action. For example, DHS reports the number of "high—risk" facilities has already been reduced by more than 2,000.

It's important that we don't lose this momentum. Dow and ACC support the permanent reauthorization of CFATS because we believe it is a successful program that meets the regulatory principles outlined by the President. The program has laid a solid foundation to address security at chemical facilities while at the same time providing industry the flexibility it needs to remain competitive and provide the innovative products that are vital to the nation's economy.

We believe any changes to the program that would require implementation or even enhanced scrutiny of any single risk-reduction option that is available to industry, such as the mandatory consideration or implementation of inherently safer technology (or "IST"), are unnecessary. The program is predicated on the idea that performance standards are the regulatory tool of choice for ensuring security, as they can be met by the facility selecting from a variety of layered security measures that best suit that specific facility. Allowing for the imposition of a single type of risk-mitigation measure would only serve to undermine the success of the program to date and flies in the face of the goal of the administration to establish the "best, most innovative and least burdensome tools for achieving regulatory ends." Such a change at this critical juncture would add an unacceptable degree of uncertainty to a program that is maturing, and place an unjustifiably large burden on chemical facilities and DHS. Accordingly, Dow and ACC do not support mandatory consideration or mandatory implementation of IST or any other single risk reduction tool.

As I have outlined in my testimony, Dow along with the members of the ACC are committed to continuing an aggressive approach in safeguarding America's chemical facilities. It is in this spirit that we look forward to working alongside DHS and this committee. Thank you for this opportunity and I am happy to answer any questions at the appropriate time.

In: Chemical Facility Security
Editors: M. J. Flores et al.
ISBN 978-1-61942-925-3

Chapter 5

CHEMICAL FACILITY SECURITY HEARING—MANNAN TESTIMONY*

M. Sam Mannan

INTRODUCTION

My name is M. Sam Mannan and I hold a BS, MS, and PhD in chemical engineering. I am a registered professional engineer in the states of Louisiana and Texas and I am a certified safety professional. I am a Fellow of the American Institute of Chemical Engineers and a member of the American Society of Safety Engineers, the International Institute of Ammonia Refrigeration, and the National Fire Protection Association. I am Director of the Mary Kay O'Connor Process Safety Center, holder of the T. Michael O'Connor Chair I in Chemical Engineering, and Regents Professor of Chemical Engineering at Texas A&M University. The center seeks to develop safer processes, equipment, procedures, and management strategies that will minimize losses in the process industry. My area of expertise within the chemical engineering discipline is process safety. I teach process safety engineering both at the undergraduate and graduate level. I also teach continuing education courses on process safety and other specialty process

---

* These remarks were given as testimony on February 11, 2011. M. Sam Mannan is Regents Professor and Director of the Artie McFerrin Department of Chemical Engineering MS-3122 at Texas A&M University..

safety courses in the United States and overseas. My research and practice is primarily in the area of process safety and related subjects. The opinions presented in this document represent my personal position on these issues. These opinions are based on my education, experience, research and training.

Chemical security and protection of the chemical infrastructure is of extreme importance to our nation, and I am pleased that the US Congress is continuing to pay attention to issues relating to Chemical Facility Anti-Terrorism. I have provided testimony previously on this subject to the US House of Representatives Subcommittee on Transportation Security and Infrastructure Protection of the Committee on Homeland Security, on December 12, 2007 and the US Senate Homeland Security and Governmental Affairs Committee on March 3, 2010. Since then, I have continued to study various issues related to inherent safety and the implementation of inherent safety. While we have gained additional insight about inherent safety issues, my opinions remain much the same as then.

BACKGROUND

Hazardous materials can be grouped into three tiers of vulnerability categories. The first category includes the stationary facilities that are members of major industry associations. Even though these facilities have large inventories of hazardous materials and are quite visible, they are the best prepared against attack because of voluntary programs that have been developed and implemented. The second tier of vulnerability category includes smaller and medium-sized facilities that manufacture or use chemicals but may or may not be members of any industry associations. These facilities are less visible, but are also, in general, less prepared and more widely distributed. Finally, the third category of vulnerability includes all hazardous materials that are in transit (by whatever means) throughout the United States. In addition to being present almost anywhere in the United States at any given time, this category also represents high visibility and the highest vulnerability. It could also be argued that this category is the least prepared to deal with intentionally caused catastrophic scenarios.

Some pertinent subjects of interest with regard to attacks on the chemical infrastructure are: active protection measures; passive protection measures; vulnerability analyses, response and recovery plans; and long-term needs and priorities. Active protection measures include increased security, limited access to facilities, and background checks. Examples of passive protection measures include development of exclusion areas and process and engineering measures.

Vulnerability analysis, response, and recovery plans are needed not only to help devise the prevention and protection plans, but also to develop the response and recovery plans. In this respect, it must be mentioned that most of the large, multi-national facilities that are members of major industry associations have voluntarily conducted some form of vulnerability analysis. What is not clear is whether these analyses have been used to integrate planning for response and recovery efforts in coordination with local agencies and the public. One very stark lesson from the 9/11 events is that the “first” first-responders are usually members of the public. Additionally, area- and region-specific vulnerability analysis and assessment of infrastructure availability for response and recovery have not been conducted. Finally, a national vulnerability analysis and assessment of infrastructure availability for response and recovery is a critical need.

Whether natural or man-made, disasters will continue to happen. However, as we have seen with the 9/11 events, hurricanes Katrina and Rita, and chemical incidents such as the Bhopal disaster, planning and response is crucial in being able to reduce the consequences and to recover from the disaster more rapidly. In this regard, it is essential to conduct vulnerability analysis, response, and recovery planning at the following three levels:

- **Plant-specific vulnerability analysis** and assessment of infrastructure availability and preparedness for response and recovery is needed. As mentioned earlier, most of the large multi-national facilities that belong to prominent industry associations have voluntarily conducted some form of vulnerability analysis. What is not clear is whether these analyses have been used to integrate planning for response and recovery efforts in coordination with local agencies and the public.
- **Area- and region-specific vulnerability analysis** and assessment of infrastructure availability for response and recovery should be conducted. Each area- and region-specific analysis should include an assessment and planning for evacuation and shelters.
- **National vulnerability analysis** and assessment of infrastructure availability for response and recovery is critically needed. In doing this national analysis, impact on international issues and criteria should also be considered.

Inherent safety options can be considered; however, it must be understood that determination of inherent safety options is a "*complex process*" and not an "*off-the-shelf technology.*" We must be aware of the differences in implementing inherent safety options for existing plants, as compared to new plants. Also, in some cases, a seemingly clear choice with regard to inherent safety may create some undesired and unintended consequences. Issues such as risk migration, reduction of overall risk, and practical risk reduction should be evaluated whenever an inherent safety option is considered. Such an approach should be based upon the triple-pronged philosophy: evaluation and assessment, prevention and planning, and response and recovery. Planning and preparedness is required for all three areas. Only through a comprehensive, uniform and risk-based approach can we protect the people and communities of our nation as well as protect our nation's critical chemical infrastructure.

LONG-TERM GOALS AND PRIORITIES

Long-term goals and priorities to prevent and/or reduce the consequences of intentional catastrophic scenarios require clear thinking and hard work. While no one would argue that making hazardous materials less attractive as a target should be a goal that all stakeholders should accept, differences arise in how we realize that goal.

Another long-term goal is to develop technology and know-how with regard to resilient engineered systems and terrorism-resistant plants. In this respect, research and technological advances are needed in many areas, such as bio-chemical detection, sensors, and self-healing materials. Protection of the chemical infrastructure, like many other challenges, requires the commitment and effort of all stakeholders.

I feel very strongly that science should precede regulations and standards. With regard to science and technology investments, many initiatives have been proposed and are being implemented. However, some important additional initiatives that should also be considered are given below:

1. The fact is that the chemical infrastructure and all components including the individual sites, supply, and delivery systems were never built with terrorism in mind. Research must be conducted to

determine how we might have designed and built the chemical plants and the infrastructure had we considered these threats. The ultimate goal for such research would be two-pronged. First, determine options for what can be feasibly implemented for existing plants. Second, if necessary, prescribe new standards and procedures for new plants.

2. Research investments should be made on advanced transportation risk assessment methods. Before transportation of any hazardous materials, a transportation risk assessment should be conducted using available information and methodology, as well as time-specific data that may be available.
3. Additional science and technology investments that should be considered are:
   - Development of incident databases and lessons learned. This knowledge base could then be used to improve planning, response capability, and infrastructure changes. Recent experience in this regard is the improvement in planning and response for Hurricane Rita from lessons learned from Hurricane Katrina.
   - Research should be conducted on decision-making, particularly under stress, and how management systems can be improved.
   - Research on inherent safety options and technologies. This type of research should be combined with systems life cycle analysis and review of practical risk reduction. In other words, implementation of inherent safety options should not be allowed to create other unintended consequences, risk migration, or risk accumulation. While transportation is outside the scope of the ***Chemical Facility Antiterrorism Standards (CFATS)***, it must be included in vulnerability assessments to avoid transfer of facility risks to transportation risks.
   - Basic and fundamental research is also needed on design of resilient engineered systems. For example, if the collapse of the World Trade Center towers could have been extended by any amount of time, additional lives could likely have been saved.
   - Basic and fundamental research is also needed on resilient and fail-safe control systems.
   - Long-term research is also needed in the area of self-healing materials and biomimetics.

Specific Comments Relevant to Preventing Chemical Terrorism: Building a Foundation of Security at Our Nation's Chemical Facilities

With regard to the *Chemical Facility Antiterrorism Standards*, I have the following specific comments:

1. The US Congress must give the Department of Homeland Security permanent and continuing authority to regulate chemical security in the United States. While many facilities are voluntarily taking appropriate measures, I am concerned that many are not. A regulation that creates a minimum and level playing field is very important.
2. The use of a risk-based approach and risk-tiering in evaluating the vulnerability of any facility is a good approach.
3. The Department of Homeland Security does not currently have appropriate and adequate expertise to implement and enforce inherent safety.
4. With regard to mandated consideration and/or implementation of inherent safety in chemical facility antiterrorism regulations, I have the following comments:
    a. Despite a significant amount of work in the area of inherent safety design and technology, it still remains a very complex task to determine on a case-to-case basis what is an inherently safer technology or approach. In many cases, a seemingly inherent safety approach may result in unintended consequences, risk transfer, and/or risk accumulation.
    b. Based on current know-how and science, there does not exist any widely accepted scientific process by which to require (by legislation or regulation) a mandatory assessment of "inherently safer technology," at a chemical facility. As a result, there are dangers associated with mandating a specific assessment model or requiring an overly burdensome assessment regime.
    c. There are many methods available to the industry for potentially reducing risk and vulnerability. Vulnerability assessments should consider the feasibility of all methods for improving security to determine the method to achieve the optimum balance of cost effectiveness and vulnerability reduction.
    d. As I stated earlier, science should precede regulations. I do not believe that the science currently exists to quantify inherent

safety. Regulations or any actions taken as a result of regulations should not create unintended and unwanted consequences. An example in this context is the substitution of hydrogen fluoride (HF) with sulfuric acid (H_2SO_4) for refinery alkylation processes. While it is true that HF is more toxic than H_2SO_4, the amount of H_2SO_4 needed to do the same amount of processing is 100-140 times or more than HF. Thus changing from HF to H_2SO_4 would require large storage facilities and more transportation. In fact, changing from HF to H_2SO_4 may provide more opportunities for a terrorist attack. On the other hand, a well- managed plant with a smaller amount of HF and appropriate safety protective systems may represent a lower overall risk.

e. An example of risk transfer as well as risk accumulation is the replacement of chlorine with sodium hypochlorite for water treatment processes. The sodium hypochlorite itself is manufactured from chlorine and thus the risk is transferred somewhere else. In fact, if all water treatment is converted to sodium hypochlorite processes, it would lead to risk accumulation where a mega-plant would have to be constructed somewhere with large quantities of chlorine representing a high-value target. Thus, while conversion to sodium hypochlorite may be advisable in some cases, but not in all cases. The determination should be made on a case-by-case basis depending on case- specific information and life cycle risk analysis.

f. In the case of conversion from chlorine to sodium hypochlorite, there could also be unintended consequences. Sodium hypochlorite is known to be unstable and can decompose into chlorates which subsequently convert to perchlorates. Depending on the duration and conditions of storage of sodium hypochlorite, the formation of perchlorates can occur. This presents an obstacle when this alternative is introduced as an inherently safer technology than that of chlorine since perchlorates are toxic and have been noted to induce thyroid problems from extended exposure. Moreover, perchlorates have been found in more than 90% of sodium hypochlorite samples, and its concentration is believed to increase with age. Contamination of drinking water during its treatment is the most likely way of entering the human body. However, it has also been found that plants can absorb perchlorates without adverse effects to themselves. The perchlorates are then stored may be passed along in the food chain[1].

A study performed by Orica (an Australian company) found that the number of incidents involving chlorine gas formed by the inadvertent mixing of acid with sodium hypochlorite was larger than those directly from chlorine gas alone[2]. This fact, along with the hazardous properties of the chlorates formed by improper storage of sodium hypochlorite shows that this option cannot be readily listed as an inherently safer alternative for chlorine and that its risks cannot be overlooked.

g. When inherent safety options are considered, we must understand and account for the challenges and difficulties in implementing inherently safer technology and options. The first challenge is simply to measure the degree of inherent safety in a way that allows comparisons of alternative designs, which may or may not increase safety or may simply redistribute the risk. The second is that because inherent safety is an intrinsic feature of the design, it is best implemented early in the design of a process plant, while the US has a huge base of installed process plants and little new construction. Finally, in developing inherently safer technologies, there are significant technical challenges that require research and development efforts. These challenges make regulation of inherent safety very difficult. We believe that a coordinated long-term research and development effort involving government, industry, and academia is essential to develop and implement inherently safer technologies. A similar collaborative approach has shown success in related areas such as green chemistry, energy conservation, and sustainable development.

h. Instead of prescriptive requirements for inherently safer technology and approaches, facilities should be allowed the flexibility of achieving a manageable level of risk using a combination of safety and security options. For example, nuclear facilities have very high hazard materials, but they protect their site and the public with a combination of multiple layers of security and safety protective features. All methods of reducing vulnerability should be considered on a case-by-case basis, and the implementation of any one particular method should not take or appear to take precedence over the others.

i. Over the past 15-20 years, and more so after 9/11, consideration of Inherently Safer Technology (IST) options and approaches has effectively become part of industry standards, with the experts and persons with know-how assessing and implementing inherently safer options, without prescriptive regulations that carry risks (both as trumping other tools or potentially shifting risk). A better approach for applying IST in security is by allowing the companies to assess IST as part of their overall safety, security and environmental operations and therefore, cannot be prescriptive. In fact, it seems that the current risk tiering of stationary sources inherently encourages the use of IST. For example, a facility that is in Tier 1 would consider all IST options if they could move the facility to Tier 3 or Tier 4. So, it could be argued that consideration of IST options is part of the market-driven incentive-based approach within the current CFATS legislation. In fact, when facilities use IST approaches to change their tier designation, they should be required to prove that:
   - Overall risk has been reduced
   - That unintended consequences have not been created
   - That risk transfer and/or risk accumulation has not occurred
   - That the actions taken or proposed have not just simply resulted in the departure of industry from one area to another

5. Cost-benefit analysis of inherent safety options is another very important and pertinent issue that should be a significant part of any decision making process.
6. Currently there does not exist a generally accepted understanding on the definition of "inherently safer technology." Given that background, if regulations require all plants to be "inherently safe," there might be a tendency to broaden the definition of "inherently safe," so that almost everything fits the definition.
7. Before adopting any regulatory framework requiring the evaluation and/or implementation of inherently safer technology, significant research questions must be answered to reach a universally accepted definition of "inherently safer technology." Research in critical areas such as system reliability and resilience must also provide information

to help develop appropriate guidance for facilities, both new and old, regarding methods to assess the costs, benefits, and potential risks of process changes at their facilities and throughout the supply chain and market.

SUMMARY

I applaud the US Congress for providing leadership in this important area of chemical security. It is clear that many companies are taking reasonable and responsible steps in chemical security, including the consideration of inherent safety options. Inherently safer technology is an objective that should continually be pursued, but must always be based upon sound science as well as sound risk assessment and management principles. Mandating the evaluation and/or implementation of inherent safety options must be based on good science. Before such steps can be taken, important issues must be resolved such as a generally accepted understanding on the definition of "inherently safer technology," methods for quantification of inherent safety, and methods for evaluation of inherent safety options. I do not believe that current know-how and science exists to adequately define and quantify any of these issues.

Requirements for inherently safer technology should be based upon good science aimed at making the industry secure, avoid over-regulation, and create a level playing field. US facilities could be at a competitive disadvantage if required to implement unproven technologies simply to meet a regulator's position that such technology is more inherently safe.

I am encouraged by the leadership of Congress and by continued efforts to seek expertise and opinion from all stakeholders.

End Notes

[1] http://www.watertechonline.com/news.asp?N_ID=70952
[2] http://www.wioa.org.au/conference_papers/04/paper2.htm

In: Chemical Facility Security
Editors: M. J. Flores et al.
ISBN 978-1-61942-925-3

Chapter 6

CHEMICAL FACILITY SECURITY HEARING—HAWKINS TESTIMONY[*]

George S. Hawkins

Good afternoon Chairman Lungren, Ranking Member Clarke and members of the Subcommittee on Cybersecurity, Infrastructure Protection and Security Technologies. My name is George Hawkins and I am the General Manager of the District of Columbia Water and Sewer Authority – also known simply as DC Water. I'd like to thank you for inviting me to testify today on the circumstances surrounding the decision to have the Blue Plains Wastewater Treatment Facility voluntarily switch from using a chlorine in the treatment of wastewater to a potentially safer alternative.

INTRODUCTION

First, by way of background, DC Water purchases treated drinking water at wholesale from our federal partner, the Washington Aqueduct, which disinfects our drinking water supply and is a unit of the U.S. Army Corps of Engineers. We then deliver this water through our pumping stations and pipes to our retail customers in the District of Columbia – including this very building. We also operate the world's largest advanced wastewater treatment

---

[*] These remarks were given as testimony on February 11, 2011. George S. Hawkins is General Manager for the District of Columbia Water and Sewer Authority.

plant, at Blue Plains, for the benefit of our customers in the District and several suburban jurisdictions. We serve more than two million customers in the metropolitan Washington, D.C., area. The disinfection of wastewater provides critical public health protection. Disinfection destroys bacteria and viruses, helping to protect ecosystems and prevent waterborne disease. The most commonly used disinfectant for both drinking water and wastewater treatment is chlorine. Its effectiveness against a wide spectrum of disease causing organisms, its relatively low cost, and high reliability contribute to its popularity. Chlorine can be applied to water directly as a gas, or through the use of chlorinating chemicals. A number of alternative disinfectants, such as chlorine dioxide, chloramines, ozone, and ultraviolet radiation, are also used to varying degrees. Each disinfection technology has unique benefits, limitations, and costs. Individual water system operators must weigh these trade-offs and choose disinfection methods based on local water quality conditions, climate, physical limitation of plant location, cost, compliance with the Clean Water Act (CWA) and the needs and resources of the communities they serve. Based on this wide variety of factors, use of alternative chemicals may not be possible for all wastewater utilities.

My colleagues throughout the water sector are currently examining this issue very closely, not only to protect the populations they serve but also to protect their most critical asset-their workforce. In 2009, the National Association of Clean Water Agencies (NACWA) conducted an informal survey of its membership which shows that clean water agencies are using other treatment technologies when local factors enable them to do so. In fact, 66% of survey respondents indicated they no longer use gaseous chlorine in their disinfection process. Of the 33% that continue to utilize gaseous chlorine, 20% planned to switch to another disinfectant within a one to two year timeframe. Keeping in mind that NACWA members account for about 80% of the treated sewerage stream in the United States you can see utilities switching to other treatments when possible. Given this information we do not believe a federal mandate is necessary, rather we believe that decisions regarding treatment technologies should reside within the local community.

Blue Plains

The Blue Plains Wastewater Treatment Facility sits on the Potomac River in southwest Washington, D.C. Like most facilities before 9-11, Blue Plains used chlorine gas and other hazardous compounds in its treatment process. In fact, when the plane hit the Pentagon on 9-11, Blue Plains had six 90-ton railcars on site storing dangerous chemicals just four miles away from the Capitol. Three were filled with chlorine gas, one was filled with sulfur dioxide, and two railcars were being unloaded; each filled with chlorine gas and sulfur dioxide.

Not only were these chemicals a threat when in storage on the site, but they also created a hazard in transit as they were hauled by rail through downtown D.C. on their way to Blue Plains. To get a sense of the exposure they represented, consider that in January 2005, when a freight train pulling three tankers full of liquidized chlorine and one tanker of sodium hydroxide slammed into a parked train in Graniteville, South Carolina, it released 11,500 gallons of chlorine gas. Nine people died and at least 529 were injured. That was without any malicious intent and in a rural location.

Chlorine gas was infamous during trench warfare in World War I. It irritates the eyes, nose, and skin in small amounts, but in concentrated form, the yellowish green gas causes the lungs to fill with liquid, drowning the victims to death.

Prior to 9-11, personnel at Blue Plains were concerned over the hazardous nature of the disinfecting chemicals and the potential for accidental release. A Capital Project was introduced into the 10 year Capital Planning Cycle to replace chlorine and sulfur dioxide gases with safer liquid chemicals such as sodium hypochlorite (bleach) and sodium bisulfite. Liquid bleach is much safer than chlorine gas, being 6 percent stronger than household bleach and easy to contain if spilled. After 9-11, facility personnel at Blue Plains were concerned over the prospect of terrorism. We evaluated our situation and decided it would make more sense for our location in the nation's capital to fast track the switch over to liquid chemicals that are much safer. The solution was to switch to using sodium hypochlorite for the purpose of disinfection and to the use of liquid sodium bisulfite for de-chlorination of residual chlorine in the waters being discharged to the river. A Process Safety Management (PSM) Committee was formed, including safety, engineering, facilities and maintenance personnel to ensure that chlorination and dechlorination systems were safe as possible until they were finally decommissioned. We completed this conversion in two phases. In the first phase, a temporary liquid disinfection/dechlorination system was designed, installed, and made operational within 60 days of 9-11 allowing removal of the chemical rail cars from site. We simultaneously fast tracked the Capital Project for the installation of the permanent liquid disinfection/dechlorination system.

Costs

As discussed, we first built a temporary facility and purchased additional storage tanks for the liquid bleach and bisulfite, as well as pumps and piping to deliver the chemicals to the wastewater in the right dose and at the correct locations while we pursued construction of the more capital intensive permanent conversion. By October 2003, we had finished the permanent conversion to our plant. The process of converting the old plant was costly; it required adding more storage tanks, pumps, piping and instrumentation than had been needed before. We also had to build additional storage facilities for the liquid bleach and sodium bisulfate-used for dechlorination. The total cost was $16.4 million, including the installation of the temporary facilities.

Operating costs are also now higher as well. The driving factor is that liquid bleach is much more expensive than chlorine gas. The annual cost of purchasing chemicals has increased from approximately $800,000 annually for gas chlorine to over $2 million annual average for sodium hypochlorite.

In our case, the switch effectively removed the threat of harmful exposure for 1.7 million people living near the Blue Plains plant. There is no longer any risk to the public since the conversion. Second, the switch from chlorine gas also simplified the plant's operations in several ways. For example, because liquid bleach is much safer to handle, the switch has limited the amount of training that employees need and reduced accidents. Last, the threat of a terrorist attack has diminished. At one time we routinely stored five or six tankers on site. Had a catastrophic leak occurred, this could have caused many fatalities and injuries within a 10 mile radius.

Regulation

Since the enactment of the Public Health Protection and Bioterrorism Preparedness and Response Act of 2002, the Environmental Protection Agency has regulated the physical security of the nation's drinking water systems through Section 1433 of the Safe Drinking Water Act (SDWA). Under this law, drinking water systems serving more than 3,300 people were required to prepare vulnerability assessments detailing risks related to possible terrorist attacks, and emergency response plans outlining procedures for responding for such an attack. EPA has reported that virtually all covered drinking water systems are in compliance with these requirements, and in 2006

Congress exempted the water sector from additional physical security regulation through the Department of Homeland Security's Chemical Facilities Anti-Terrorism Standards (CFATS).

Given the importance of coordinating drinking water security rules with the public health requirements of SDWA, most drinking water systems believe that their exemption from CFATS is appropriate, and that EPA should continue oversight of any new or revised drinking water security program. Similarly, any federal security regulations imposed on the wastewater and water sector should come through a comparable EPA program rather than CFATS. A regulatory approach that were to divide drinking water and wastewater security among different federal agencies could lead to confusing and contradictory standards – especially for utilities that provide both drinking water and wastewater service to a community.

The Obama Administration has gone on record in support of ensuring water and wastewater systems are regulated by the Environmental Protection Agency and not the Department of Homeland Security, testifying in 2009, "EPA should be the lead agency for chemical security for both drinking water and wastewater systems." Chairman Lungren, members of the subcommittee, this concludes my prepared remarks. Thank you again for the opportunity to testify, and I look forward to answering any questions you may have.

In: Chemical Facility Security
Editors: M. J. Flores et al.
ISBN 978-1-61942-925-3

Chapter 7

CHEMICAL SECURITY HEARING—SILVA TESTIMONY*

Peter S. Silva

INTRODUCTION

Good morning Mr. Chairman, Ranking Member Collins, and Members of the committee. I am Peter Silva, Assistant Administrator for Water at the United States Environmental Protection Agency. I welcome this opportunity to discuss EPA's efforts to promote security and resiliency in the Water Sector with an emphasis on our role in addressing chemical security at water facilities.

EPA has worked over the last several years to support the Water Sector in improving security and resiliency, and I am pleased to report that the sector has taken its charge seriously. EPA has been entrusted with important responsibilities for coordinating the protection of the Water Sector through Congressional authorization under the *Public Health Security and Bioterrorism Preparedness and Response Act of 2002* (the Bioterrorism Act), and through presidential mandates under Homeland Security Presidential Directives (HSPD) 7, 9 and 10. HSPD 22 (the domestic chemical defense) also applies to water protection.

* These remarks were given as testimony on March 3, 2010. Peter S. Silva is Assistant Administrator for Water at the US Environmental Protection Agency.

Promoting the security and preparedness of the nation's water infrastructure remains a priority of the agency in a post-9/11 and post-hurricane Katrina world. A loss of water service can seriously jeopardize the public health, economic vitality, and general viability of a community. In working with the Water Sector, we have emphasized a multi-layered approach to security consisting of prevention, detection, response, and recovery so that we can assist water facilities in avoiding incidents and, should an incident occur, in quickly identifying and recovering from such events.

IMPLEMENTATION OF SECTION 1433 OF THE SAFE DRINKING WATER ACT

Existing statutory requirements address chemical security at drinking water systems to a degree. Section 1433 of the Safe Drinking Water Act (added by the Bioterrorism Act of 2002) required each community water system providing drinking water to more than 3,300 persons to conduct a vulnerability assessment, certify its completion, and submit a copy of the assessment to EPA. These vulnerability assessments addressed security at water systems comprehensively, from water collection to treatment and distribution, and they specifically included the use, storage, or handling of chemicals. In addition, Section 1433 required each water system to prepare or revise an emergency response plan that incorporates the findings of the vulnerability assessment and to certify to EPA that the system has completed such a plan.

Since 2003, EPA has received 100% of the vulnerability assessments and emergency response plan certifications from large and medium community water systems serving more than 50, 000 people. Over 99% of small community water systems serving between 3,300 and 50,000 people have submitted their vulnerability assessments and emergency response plan certifications. EPA reviewed the vulnerability assessments to ensure compliance with Bioterrorism Act, and where necessary provided technical assistance to individual drinking water systems to bring these systems into compliance. EPA also initiated some enforcement actions against non-compliant systems.

EPA's Role in Chemical Security for Drinking Water and Wastewater Utilities

EPA's current approach for addressing chemical security in the Water Sector involves a longstanding effort to promote the voluntary adoption of countermeasures by water facilities. Before I discuss some of these activities, however, I would like to take a step back to consider the broader implications of chemical security for the Water Sector. It is of paramount importance for us to acknowledge in this discussion that the primary purpose of drinking water systems is the provision of safe drinking water, while that of wastewater systems is the protection of water bodies. In fact, the effective treatment of drinking water to control infectious diseases like typhoid and cholera has been hailed by the U.S. Centers for Disease Control and Prevention as one of the greatest public health achievements of the twentieth century.

Therefore, authorizing language should allow for a consideration of this essential public health and environmental mission, particularly with respect to any provision which may require a facility to consider alternative water treatment processes. In other words, chemical security regulations when applied to the water sector should enable a reasoned balance of multiple, important factors so that we can achieve the joint policy goals of protecting public health and the environment while enhancing security. Such factors include: efficacy of treatment in meeting public health and environmental requirements, security concerns, reliability of treatment, source water characteristics, feasibility, and operator safety.

Tools and Technical Assistance

EPA has worked closely with the water sector to assess and reduce the risks associated with hazardous chemicals. To this end, EPA and industry associations, often in partnership, have developed tools, training and technical assistance to help water utilities identify and mitigate those risks. A few examples of our efforts are as follows:

1. We developed tools that assist water systems with assessing vulnerabilities, including chemical storage and handling. Examples of the tools include:
   - The *Vulnerability Self Assessment Tool (VSA TTM)* – a software package that supports water and wastewater utility vulnerability assessments using a qualitative risk assessment methodology;
   - The *Security Vulnerability Self-Assessment Guide for Small Drinking Water Systems* – a manual specifically designed to help small water systems conduct vulnerability assessments; and
   - The *Security Vulnerability Self-Assessment Guide for Very Small (<3,300) Systems*, which assists these systems in assessing their critical components and identifying security measures that should be implemented.
2. Under the Bioterrorism Act of 2002, EPA created a document to "provide baseline information to community water systems... regarding which kinds of terrorist attacks or other intentional acts are the probable threats to: (A) substantially disrupt the ability of the system to provide a safe and reliable supply of drinking water; or (B) otherwise present significant public health concerns." The baseline threat document addressed vulnerabilities related to the use, transfer and storage of chemicals, including the evaluation of different disinfection options. EPA provided this document to drinking water facilities to assist them in conducting their vulnerability assessments.
3. The National Association of Clean Water Agencies (NACWA) has worked with the Department of Homeland Security (DHS) and EPA to create a Chlorine Gas Decision Tool for Water and Wastewater Utilities. The Tool is designed to provide utilities with a user-friendly, but thorough, means of evaluating alternatives to chlorine gas disinfection.
4. EPA created a series of Security Product Guides that assist water facilities with making enhancements to reduce risks and protect against man-made and naturally occurring events. These guides provide recommendations for improving physical security, such as the use of barriers, placement and security of aboveground equipment, selection of fencing materials, and the use of visual surveillance monitoring systems, all of which can help to secure hazardous chemicals used by water facilities.
5. We funded a cooperative agreement with the American Society of Civil Engineers, the American Water Works Association, and the

Water Environmental Federation to develop Voluntary Physical Security Standards for drinking water and wastewater systems. Completed in December 2006, these voluntary standards address storage of hazardous or toxic chemicals, including chlorine and ammonia gas.

6. Together with the National Oceanic and Atmospheric Administration, EPA developed ALOHA (Aerial Locations of Hazardous Atmospheres) and RMP*Comp – software tools that many water utilities and other facilities use to model the dispersion of hazardous substances. DHS uses RMP*Comp in its Chemical Facilities Anti-Terrorism Standards (CFATS) program.

Risk Management Plans

In addition to the above activities, EPA's Chemical Accident Prevention Provisions (40 CFR 68.1 - .220), developed under the authority of the Clean Air Act, Section 112(r), require utility processes containing certain levels of specific hazardous substances to implement an accident prevention program, conduct a hazard assessment, prepare and implement an emergency response plan, and submit to EPA a summary report known as a risk management plan (RMP). The RMP must describe the facility's accidental release prevention and emergency response policies, the regulated substances handled at the facility, the worst-case release scenario(s) and alternative release scenario(s), the 5-year accident history of the facility, the Emergency Response Plan, and planned changes to improve safety at the facility (see 40 CFR Part 68). Hazardous chemicals of most relevance to the Water Sector, including gaseous chlorine, ammonia, sulfur dioxide, and chlorine dioxide, trigger RMP regulatory requirements if they exceed certain threshold quantities.

Considerations for Chemical Security in the Water Sector

It is important to note that the administration has developed a set of guiding principles for the reauthorization of the Chemical Facility Anti-Terrorism Standards (CFATS) program and for addressing the chemical security of our nation's wastewater and drinking water treatment facilities. These principles are:

1) The administration supports permanent chemical facility security authorities and a detailed and deliberate process for doing so, hence our preference for that process to be completed in FY1 0.
2) Nonetheless, CFATS single year reauthorization in this session presents an opportunity to promote the consideration and adoption of inherently safer technologies (IST) among high risk chemical facilities. We look forward to working with this committee and others on this important matter.
3) CFATS reauthorization also presents an opportunity to close the existing security gap for wastewater and drinking water treatment facilities by addressing the statutory exemption of these facilities from CFATS. The administration supports closing this gap.

As we have stated to Congress before, we believe that there is a critical gap in the U.S. chemical security regulatory framework—namely, the exemption of drinking water and wastewater treatment facilities. We need to work with Congress to close this gap in order to secure substances of concern at these facilities and to protect the communities they serve. Drinking water and wastewater treatment facilities that meet CFATS thresholds for chemicals of interest should be regulated. We do, however, recognize the unique public health and environmental requirements and responsibilities of such facilities. For example, we understand that a "cease operations" order that might be appropriate for another facility under CFATS would have significant public health and environmental consequences when applied to a water facility.

The administration also has established the following policy principles in regards to regulating security at water sector facilities:

- EPA should be the lead agency for chemical security for both drinking water and wastewater systems, with DHS supporting EPA's efforts.
- To address chemical security in the water sector, EPA would utilize, with modifications as appropriate given statutory requirements and the uniqueness of the sector, DHS' existing risk assessment tools and performance standards for chemical facilities.
- DHS should be responsible for ensuring consistency of high-risk chemical facility security across all 18 critical infrastructure sectors.
- Where possible, we support using safer technology, such as less toxic chemicals, to enhance the security of the nation's high-risk chemical storing facilities. Facilities posing the highest degree of risk should be required to implement IST method(s) if such methods enhance overall

security, are feasible, and consider public health and environmental requirements.

In addition to articulating these principles, I would like to present the agency's view of the appropriate division of regulatory labor between EPA and the states. EPA supports legislative authority which would provide the states with an important role in regulating chemical security at water systems, including a prominent state role in IST determinations and auditing/ inspections. This federal partnership with the states would leverage long established EPA-state relationships under the drinking water and wastewater programs, as well as the states' expertise and familiarity with individual water facilities.

CONCLUSION

Over the past several years, we have made progress in ensuring the security of our nation's drinking water and wastewater systems. We have produced a broad array of tools and assistance that the Water Sector is using to assess its vulnerabilities, reduce risk, and prepare for emergencies, including chemical theft and release. In developing these tools, we have worked effectively with our partners within the sector, and also reached out to build new relationships beyond the sector, to ensure that water utilities can be prepared to prevent, detect, respond to and recover from intentional incidents and natural disasters.

With respect to security at water sector facilities, we look forward to continuing to work with members of the committee on legislation that ensures the security of drinking water and wastewater facilities while supporting the critical mission of these facilities for public health and environmental protection.

Thank you again for the opportunity to testify about our role in water security. I would be happy to answer any questions you may have.

INDEX

D

E

F

G

H

I

J

L

M

N

O

P

Q

R

S

T

U

V

W